ROBIN S

Tread Lightly

A Guide to Travelling Green in Australia

Black Inc.

Other books by Robin include:

The Clean House Effect: Hundreds of Practical, Inexpensive Ways To Reduce the Use of Chemicals in the Home

Moonbird, a novel for teenagers

New Faces: The Complete Book of Alternative Pets (CBC Book of the Year 1995)

From Seeds to Leaves: A Complete Guide to Growing Australian Trees and Shrubs from Seed

Envirocat: A New Approach to Caring for Your Cat and Protecting Native Wildlife

The Dog Book: How to Choose a Dog That Suits Your Personality and Lifestyle

Wombat and Koala: Bush Babies Solo Series

Alternative Pets: From Budgies and Yabbies to Rabbits and Rats

Robin Stewart's Chemical-Free Home: Hundreds of Practical and Inexpensive Ways to Reduce the Use of Chemicals in Your Home

Robin Stewart's Chemical-Free Pest Control: Hundreds of Practical and Inexpensive Ways to Control Pests Without Chemicals

Australian Green Home and Garden

Charles Darwin's Big Idea

Dedication

For people who recognise that we're part of the environment, rather than separate from it.

Acknowledgements

My husband, Doug, has for the past 36 years, shared my passion for all things natural and the environment. Together we've made this journey of discovery – created this book.

Sincere thanks to my friends and relatives who've given so generously of their time and hints. Also, to all the authors of books I've absorbed over the years.

Lastly, I'd like to thank my editor, Eugenie Baulch, for her editorial expertise. Also, the entire team at Black Inc. who are always a pleasure to work with.

Published by Black Inc.

Level 5, 289 Flinders Lane
Melbourne VIC 3000
email: enquiries@blackincbooks.com
http://www.schwartzpublishing.com

National Library of Australia Cataloguing-in-publication data:

> Stewart, Robin E. (Robin Elaine), 1943- .
>
> Tread lightly : a guide to travelling green in Australia.
>
> ISBN 1 86395 349 3.
>
> 1. Tourism - Environmental aspects - Australia. 2.
>
> Australia - Description and travel. I. Title.
>
> 919.4047

Publisher's Note

The hints and remedies contained within these pages are suggestions only.

It is always wise to experiment, prior to application on a larger
scale, due to the wide variation of materials used in the manufacturing process.

Every effort has been made to ensure that this book is free from
error or omissions. However, the publisher, the author, the editor,
or their respective employees or agents, shall not accept responsibility for injury, loss, or damage occasioned to any person acting
or refraining from action as a result of material in this book
whether or not such injury, loss, damage is in any way due to any
negligent act or omission, breach of duty or default on the part of
the publisher, the author, the editor or their respective employees
or agents.

Printed in Australia by Griffin Press.

Contents

Introduction: Sustainable Travel

Every winter, caravans, trailers and motorhomes head north in their thousands, flocking like migratory birds to Queensland's Gold Coast and Sunshine Coast, then travelling on to the Centre and Kakadu. Escaping the chill and cobwebs of winter. Experiencing the feel of sunshine on bare legs, gazing sleepily into the flames of a campfire, smelling the scent of crushed gum leaves, seeing flashes of colour as rosellas swoop through the trees; gazing in awe at Uluru.

Over the years I've watched with dismay as people become more and more detached from their environment. The littering and fouling of beautiful places fills me with disappointment and shame for our species. Times are changing, however, and these same people are now beginning to be more environmentally aware; are starting to think green.

Loving Australia to death is a very real possibility. The burning of anything and everything flammable in campfires, tyre tracks gouged into sand dunes, litter cast aside carelessly and human faeces polluting river systems – when multiplied hundreds of thousands of times, these things damage Australia's fragile ecosystems. We must learn to enjoy the outdoors without endangering wildlife or damaging the very structure on which it depends. Our Australian environment is more delicate than it seems.

A fine balance exists between the natural world and the earth on which we tread. The preservation of large areas of natural habitat is, therefore, the key to the survival of indigenous plants and animals. Wilderness reserves in the form of national parks, state forests and World Heritage areas are of prime importance. After all, as Aboriginal cultures learned long ago, if you don't respect your environment, you die. The interdependence between the land and humans – of all cultures and origins – is as simple as that.

For a variety of reasons, including the War on Terror, more people are choosing to travel within Australia rather than travel overseas – especially the baby boomers. I watch with interest as my peers focus on personal enjoyment and the very best of everything. We've been labelled the selfish generation, refusing to grow old gracefully, chasing youth as never before. We want to have fun, experience adventure, be free-spirited. And be dreamers as well. Travelling provides a means by which this lifestyle ambition can be fulfilled.

With more than 4 million baby boomers, Australia faces a major challenge. Will the baby boomers spoil the environment for future generations? The Flinders Ranges, Byron Bay, Lord Howe Island, Port Douglas, south-west Tasmania, the Canning Stock Route, the Birdsville Track, the Top End, the Red Centre, the Kimberleys, the Snowy Mountains and the Great Barrier Reef. The Bungle Bungle Ranges with their fragile, crumbly landscape. Will our impact on these environments prove the label the selfish generation is correct? Or will we choose instead to travel sustainably?

An enriching experience

Throughout my life I've been fortunate to experience a wide variety of travelling experiences. As children, my parents took us bush camping. As a Girl Guide, many bush and camping skills were learned. Later, as an older teenager, the challenge of hiking in the mountains – while carrying a backpack – tested and extended my resilience.

After being married for 11 months, my husband Doug and I set off around Australia and into the outback in a VW Kombi van that cost us only $100. We took our Labrador dog, and along the way rescued two orphaned kangaroo joeys that adapted well to life on the road. The journey took us just under one year. During that time we made the all-important decision to leave city life behind and 'live on the land'.

And so began a series of properties and sheep and cattle enterprises that took us all around Victoria. A converted panel van enabled us to camp in the bush or by the coast, but this set-up was not entirely successful. In 1979 we purchased land on King Island, in Bass Strait. Our sheep property had eight kilometres of spectacular ocean frontage, in the full blast of the Roaring Forties, complete with penguin and muttonbird rookeries. Life on this island paradise was hard yet satisfying. No longer did we feel the need to spend our holidays camping.

It was during this time that we made escapes of sorts, travelling north by plane to Queensland; escaping the howling gales and rain of yet another Bass Strait winter. Basking in the sunny warmth of a Queensland winter it was possible to recharge my batteries, prepare for a return to Bass Strait and the demands of an extensive sheep property hugging the south-west coast of King Island. A place littered with shipwrecks.

So the pattern of my life was laid. A restlessness arose at the beginning of May and lasted until the end of September. A desire to travel north, seeking a more relaxed lifestyle in the great outdoors while escaping the winters of southern Australia. Depending on our budget, cabin accommodation, motels, hotels, self-contained flats, units and apartments, rental houses, houseboats and on-site caravans were sampled.

People living in the northern parts of Australia, however, crave a different type of 'migration'. Travelling south to Tasmania for the months of summer provides welcome relief from extreme humidity and heat, and an escape from monsoonal downpours that start in early December and last till the end of April.

Our caravanning days began in 1999, when living on Phillip Island. Winter touring in a caravan seemed like a good idea. With its combination of mobility and comfort, we embraced a holiday lifestyle that gave us a freedom to wander, while at the same time allowing our pets to travel with us. But even this was not perfect. Our caravan didn't tow very well; it had cupboards in the wrong places; we were making do with a system that wasn't quite us.

So what did we want? A caravan that was light and towed easily, one that we could enter at lunchtime without making adjustments to the roof height, and that didn't have a fixed stove or refrigerator. It needed to be made of materials that didn't outgas harmful chemicals, to be well insulated, have a solar panel and yet not include all the glitz of commercial caravans. There was no such caravan on the market, either new or second-hand.

Therefore, Doug set himself the challenge of building a caravan designed specifically for us and our pets. His workshop – a two-car garage – underwent a massive clean-up and reorganisation. A set of plans was purchased. Recycled building materials were used wherever possible, including the door, windows, sink and other fittings from our old caravan. With welding and basic carpentry skills, along with qualifications as an engineer, Doug began work. Slowly it began to take shape. Soon it will be finished, and ready to embark on adventures of its own.

A sense of connection with the natural world

Life is rich and balanced when travelling the green way. Understanding of, respect for and sensitivity to our environment can be learned. In these modern times it's a challenge to *tread lightly*; to live in harmony with nature in its entirety.

Deep down, we know that being in unspoiled places creates feelings of tranquillity. Deep down we know we're part of the environment and yearn to be closer to it. You will find that travelling green improves the quality of your travel – it restores balance and a sense of connection.

Whatever we call ourselves – travellers, tourists, wanderers, pilgrims, gypsies – we share a love of new experiences, different lifestyles and open horizons. A certain restlessness. A desire to break loose from the usual commitments and routines, to alter the pace and meet people we'd never come across in normal everyday life. A desire to create an opportunity for reflection and a chance to find our true self.

It's all too easy to lose oneself in the everyday routines of life. When you *tread lightly* through the environment in which you travel and show respect towards local people and fellow travellers, you'll feel the full generosity of human nature as well as experience true appreciation of the natural environment.

Travelling within Australia can be sustainable. Ecologically viable tourism is the agenda.

Chapter 1
Travelling with Purpose

Charles Darwin wrote in his diary: 'If a person should ask my advice before undertaking a long voyage, my answer would depend upon his possessing a decided taste for some branch of knowledge which could by such means be acquired.'

Darwin was on the right track. There's little doubt that travelling with purpose greatly increases your mental stimulation, enjoyment and any social contacts you make along the way. Simply having a good time, or a well-earned rest, is not necessarily a recipe for successful travel. If you travel without purpose, boredom, even in the face of an ever-changing landscape, is a common outcome.

No matter who you are, there's always something to be interested in, to focus your attention on. The possibilities are endless. Keep in mind the fact that a number of interests will greatly extend and enrich your travelling experience. There's no need to restrict yourself to one passion alone.

Travellers can often obtain an excellent overview of a town, its people and surrounds from Visitor Information Centres located throughout Australia. You'll find details about special events, visitor and community information, brochures and pamphlets about accommodation and local attractions, and street maps in these centres. You'll also have the opportunity to talk to a local person, as well as buy local produce, arts and crafts. Robinvale, for example, sells locally produced honey, almonds and olives. The local people have also created beautiful crafts for sale: jewellery, pottery, wood-turning and silk-screen printing. Every town has its specialities.

Grey nomads and baby boomers commonly seek sunshine and the carefree lifestyle of their teens – before it's too late. But travel has the potential to be much more than this. Some of the following ideas may help you decide where your special interests lie.

NATURAL ENVIRONMENT

Travelling will always enrich your appreciation of nature. Some of the ways in which you can interact with the natural environment include:

- whale and dolphin watching, or swimming with dolphins, seals and whale sharks. This can be found in places as diverse as Port Phillip Bay, Monkey Mia, Bunbury and Hervey Bay
- beachcombing for cowrie shells, shark egg cases, the skeletons of sea horses, nautilus shells and so on. The greatest thrill is to find them, admire their exquisite form and then put them back for someone else to discover and enjoy in their natural setting

- collecting different types of insects, from butterflies and moths to beetles and spiders
- twitching – observing and then identifying different types of Australian birds found in the bush, on the ocean or living on inland waterways
- identifying different types of frogs, mammals, lizards and snakes
- identifying different types of orchids and wildflowers in the bush, as well as at flower shows and festivals
- studying tree species such as eucalypts and banksias
- visiting botanic gardens, as well as other significant gardens, within Australia
- fossicking or noodling for gold, opals and precious gemstones. This can be an exciting pastime, with the lure of precious stones – from topaz to sapphires to emeralds to rubies to opals to gold nuggets – a powerful incentive
- searching for fossils and different types of rock. Cut and polished agate, petrified wood or opalised shell may be your reward
- visiting some of the islands lying within Australian waters
- taking a camel tour of Uluru at sunrise or sunset, allowing you to experience Uluru in all its majesty
- discovering astronomy. Armed with your telescope, the night sky reveals its treasures – especially in the inky blackness of an outback night
- observing the breeding of migratory birds.

You can also go one step further and make a particular environmental passion the primary focus of your holiday. The Earthwatch Institute is an American organisation that offers about 130 different volunteer programs every year, world-wide. Choices range from archaeological digs to environmental activities, such as studying the echidnas on Kangaroo Island in South Australia, while at the same time having fun, meeting like-minded people and having a minimal impact on the environment. For more information, check out their website at <www.earthwatch.org>.

From Robin's travels

The steep-sided sand dune rose out of 'The Neck', a narrow isthmus of land joining North and South Bruny Islands, south of Hobart, Tasmania. Every evening at dusk from late September to early May, thousands of muttonbirds returned to their burrows after a day's fishing in the D'Entrecasteaux Channel.

A well-constructed boardwalk led to a viewing plat-form under a sky heavy with billowing black clouds. Clinging tightly to the wooden rail, with a south-south-westerly gale ripping at my parka, I climbed steadily up the 237 steps.

At the top, I gazed about me. Clothed with a dense swathe of marrum and other native grasses, along with New Zealand spinach, sword grass, pinrush and tussocks, the steep dune housed a vast underground network of muttonbird burrows. Although the dune was exposed to the harsh weather, the entrance to every burrow was well protected by thatch-like tus-socks of grass, sculpted by the prevailing wind. Inside each burrow was a male muttonbird, taking first turn at incubating the breeding pair's clutch – a single large egg.

When most of the light had faded, thousands of dark shapes materialised in the sky. A cloud of graceful dark-grey birds soared in the air currents, circling the steep-sided dune, flying fast and close together. Muted chuckles of anticipation rose from within the sandy cone as the returning birds began to crash-land among the tussocks and waddle clumsily towards their burrows. Then they were gone.

As I carefully descended the steep steps, the dune throbbed with the rich throaty coos and chuckles of thousands of muttonbirds. Their 32,000-kilometre migration flight, which sweeps north across the Pacific Ocean to the Bering Sea and back, had captured my imagination and allowed my spirit to fly free.

ADVENTURE AND PHYSICAL ACTIVITY

Many sorts of physical activities might engage you on your journey. Choose something that suits your level of fitness and health. Some possibilities include:

- yachting – either club-based activities such as touring and racing, or you can opt for life onboard, and travel without noise or pollution
- bushwalking and caving
- windsurfing, body-surfing, board-surfing, swimming, waterskiing, scuba diving and snorkelling
- exploring historic shipwrecks – there are about 6500 within Australian waters
- canoeing, kayaking (ocean, lake and river) and white-water rafting. These activities offer excitement or tranquillity, and the option of club-based activities

- cross-country skiing, bushwalking, mountaineering and rock climbing
- horse riding, travelling by camel and cycling
- motorcycle rallies
- aerobatic championships
- fruit picking. This enables you to earn-as-you-go, following the seasons, meeting up with other pickers and getting to know areas in a way that eludes most tourists
- recreational fishing (inland waterways, lake, rock, estuary, bay, offshore and surf) using rod, reel, hooks, line, sinkers, bait, lures and flies. This activity offers the chance of fresh fish for dinner and the thrill of cooking your catch. Strict fishing regulations and restrictions apply in regard to permits, bag and size limits, protected species and closed waters. These vary from state to state and from time to time
- sporting competitions such as golf, basketball and tennis tournaments, bowls carnivals and field archery championships. These create an incentive to visit a particular area, sample the natural attractions and meet people with similar interests
- prospecting with a metal detector. Who knows what will be uncovered?
- travelling to agricultural shows, dog shows, cat shows and sheep and yard dog trials
- harness racing, trotting or greyhound racing
- real estate comparisons. What would it be like to live here? These scenarios can be great fun and certainly flush out a lot of interesting information.

From Robin's travels

While travelling around Australia in the late 1960s, Doug thought he'd try surf fishing. The idea of catching our own protein was appealing. Equipped with a new rod, reel, tackle and bait, he joined a group of men lining the shore. Imitating their casting technique, Doug threw all his energy into this his very first cast. I couldn't believe my eyes. Nor could the other fishermen! Doug had cast his reel way out into the surf. As he began to pull in the full extent of his line, ten or so fishermen pretended they hadn't seen and turned away to hide their mirth. When the reel eventually emerged from the pounding waves and began its journey up the hard wet sand, the men cheered and my very red-faced husband acknowledged their gesture with a grin. 'Here, sonny,' said a white-haired fishermen. 'Take a couple of salmon back for your missus. And let me show you how to attach that reel of yours.'

ART, HISTORY AND CULTURE

You can interact with local people and explore different cultures and histories on your journey by:

- visiting local art galleries. These frequently capture the essence of a locality, increasing awareness and enjoyment of the local scenery. For example, a painting of a sea vista, with exaggerated horizontal lines, invites the traveller to look more closely into the seascapes of that particular area
- visiting country art shows, craft shops and potteries
- browsing in country museums and second-hand shops

- studying historical buildings, statues and bridges
- exploring Aboriginal art and crafts
- tracing the path of early Australian explorers
- lighthouse spotting
- tracing your genealogy and family history, with visits to country cemeteries.
- taking photographs, both underwater and on land
- giving yourself the challenge of exploring hedge mazes throughout the country
- collecting antique furniture, old books, classic motorcycles, vintage cars, teaspoons and so on
- sketching and painting, to help record the highlights of your travels. Sketches of historic buildings, bridges and so on can also be sold at local markets, helping to finance your holiday
- attending folk music, square dancing, country music, jazz, comedy and writing festivals
- visiting alternative energy expos, farmers' markets, farms, food and wine festivals and wineries, and participating in gourmet food tours
- travelling to craft festivals and shows such as those dedicated to patchwork quilting, spinning and weaving or wood turning
- visiting National Trust properties.

From Robin's travels

Tracing my family history led me to a simple bush grave located in undulating grassland near Springsure, in central Queensland. As well as visiting the cemetery, I was able to renew contact with another branch of my family.

LOCAL EVENTS

Organising your travel to take in particular events gives you access to an extraordinary diversity of interests and opportunities. Plan a trip in which you move from one exciting event to another. Don't restrict yourself to your particular interests alone. Rather, explore new ideas and broaden your horizons.

Some of the most talked-about events are included in the following lists.

Victoria

- Anglesea Angair Wildflower Festival, September
- Ararat Golden Gateway Festival, October
- Ballarat Begonia Festival, March
- Ballarat Quilters Exhibition, March
- Ballarat Antique Fair, March
- Ballarat Classic Car Rally, June
- Barmah State Park Red Gum Festival, New Year
- Bendigo Swap Meet, November. An event for old car and motorcycle enthusiasts, where precious bits and pieces are swapped or sold – recycling at its best.
- Bright Autumn Festival, April to May
- Casterton Kelpie Festival and Working Dog Auction and Kelpie Country Ball, June
- Churchill Island Bush Dance, January
- Churchill Island Working Horse Festival, Easter
- Cobram Peaches and Cream Festival, January of odd-numbered years
- Corryong High Country Festival, late March
- Daylesford Spa Festival, March
- Daylesford Highland Gathering, December
- Eildon Water Festival, October
- Euroa Wool Week, October

- Grampians Model Rail Roaders Train Exhibition, July
- Hamilton Wool Heritage Festival, late July
- Hanging Rock Races, January and March
- Hanging Rock Harvest Picnic, February
- Inverloch Jazz Festival, March
- Kilcunda Lobster Festival, January
- Mallacoota Festival of the Southern Ocean, Easter
- Melbourne Cup, November
- Melbourne Craft and Quilt Fair and Festival of Kites, July to August
- Melbourne Fringe Festival, October
- Melbourne Moomba Festival, March
- Mildura World Hot Air Balloon Championship and Country Music and Food Festival, June to July
- Mildura Country Music Festival, September to October
- Monbulk Jazz Festival, November
- Mornington Peninsula Autumn Bowls Festival, March
- Mount Macedon Autumn Floral Display, April
- Olinda Rhododendron Festival, October
- Ouyen – The Great Australian Vanilla Slice Triumph, September
- Paynesville Jazz and Swing Festival, February
- Paynesville Flat Water Classic Windsurfing Race, November
- Port Fairy Folk Festival, March
- Portland Dahlia Festival, March
- Queenscliff Music Festival, November
- Quambatook Australian Tractor Pull Championships, Easter Saturday
- Rye Beach Sand Sculpting Championships, February

- Seymour Alternative Farming Expo, March
- Seymour Rafting Festival, March
- Silvan Tulip Festival, September to October
- Sorrento Street Festival, March
- Sunbury Wine Festival, August
- Swan Hill Antique Fair, July
- Swan Hill Racing Carnival, June
- Warburton Film Festival, June
- Yarrawonga Australian Bush Poetry Championships, May

South Australia

- Adelaide Festival of Arts and Writers Festival, February
- Adelaide to Port Lincoln Yacht Race, January
- Barossa Music Festival, October
- Berri Easter Carnival, Easter
- Bordertown Camel Racing Festival, late November
- Ceduna Oyster Fest, October
- Coober Pedy Opal and Outback Festival, Easter
- Glenelg Jazz Festival, October
- Goolwa Folk and Steam Festival, mid-October
- Kingston Surf Fishing Competition, late January
- Murray Bridge Big River Challenge Festival, November
- Port Germein Festival of the Crab, early January
- Port Pirie Blessing of the Fleet, September
- Port Pirie Festival of Country Music, October
- Renmark Rose Festival, mid-October
- Robe Village Fair, November
- Victor Harbor Launch of the Whale Season, June
- Wentworth Junction Rally (vintage machinery of all types), July
- Willunga Almond Blossom Festival, July

Queensland

- Birdsville Races, September
- Bribie Island Mullet Festival, September
- Brisbane River Festival, September
- Cooktown's Endeavor Festival, June
- Fraser Island Bird Week, May
- Gatton Potato Festival, October
- Gladstone Seafood Festival, October
- Gold Coast Cup Australian Outrigger Canoe Ultra Marathon, April
- Hervey Bay Whale Festival, August
- Laidley Harvest Festival, April
- Mapleton Yarn Festival, October
- Miles Wildflower Festival, early September
- Mission Beach Banana Festival, July
- Port Douglas Reef and Rainforest Festival, early May
- Sanctuary Cove International Boat Show, May
- Toowoomba Green Week, April
- Whitsunday Food and Wine Festival and Coral Spawning, November
- Woodford Folk Festival, December

Australian Capital Territory

- Canberra Floriada Festival, Spring
- National Folk Festival, Easter

Tasmania

- Campbell Town Agricultural Show, June
- Cygnet Folk Festival, early January
- Deloraine Craft Fair, October
- Hobart Australian Wooden Boat Festival, February
- Hobart Clarence by the Water Jazz Festival, February
- Hobart Mid-Winter Festival, mid-winter

- Hobart Royal Regatta, February
- King Island Agricultural Show and Race Meeting, March
- Launceston Tamar River Festival, November
- Launceston Tasmanian Pastry Festival, early October
- Marrawah West Coast Wavesailing Classic, January
- Penguin Festival, March
- Richmond Country Music Festival, February
- St Helens Suncoast Jazz Festival, June
- Triabunna Blessing of the Fleet, October
- Ulverstone Fiesta, mid-December
- Wynyard Tulip Festival, October

Western Australia

- Albany Great Southern Wine Festival, Easter
- Albany Strawberry Fair, March
- Brookton Old Time Motor Show, March of even-numbered years
- Broome Dragon Boat Regatta, Easter
- Broome Staircase to the Moon, March
- Broome Opera Under the Stars, August–September
- Broome Pearl Festival, August–September
- Broome Mango Festival, November
- Carnarvon Festival, August
- Denmark Rainbow Festival, January
- Derby Boab Festival, late July
- Derby King Tide Day (highest tide in Australia), May
- Derby Moonrise Rock Festival, June
- Dongara Blessing of the Fleet, November
- Esperance Festival of the Wind, October–November, in even-numbered years
- Esperance Wildflower Show, September

- Exmouth Coral Coast Festival, July
- Fremantle Sardine Festival, January
- Fremantle Street Arts Festival, April
- Geraldton Batavia Celebrations (maritime history), June
- Jarrahdale Log Chop, early October
- Kalbarri Blessing of the Fleet, November
- Kalgoorlie–Boulder Great Gold Festival, July
- Kununurra Ord Festival, late August
- Marble Bar Cup and Ball, June or July
- Perth Floral and Wine Festival and Kings Park Flower Spectacular, Spring
- Rottnest Island Channel Swim, February
- Rottnest (Island) Kite Surf Classic, March
- Shark Bay Great Dingy Race, April
- Tom Price Annual Festival, early August

Northern Territory

- Alice Springs Camel Cup Carnival, July
- Borroloola Barra Classic, April
- Alice Springs Henley-on-Todd Regatta, September
- Darwin to Adelaide World Solar Challenge, September–October
- Darwin Garma Festival (Australia's most significant annual indigenous cultural event), August
- Darwin Festival (of arts and culture), September
- Jabiru Wind Festival, early September

New South Wales

- Byron Bay Writers' Festival, August
- Coffs Harbour International Buskers' Festival, September
- Deniliquin Sun Festival, late January
- Eden Whale Festival, October

- Forster Oyster Festival, October
- Gulgong Henry Lawson's Birthday Celebrations, June
- Gunnedah Eisteddfod, May
- Kangaroo Valley Country Music Festival, October
- Lismore Square Dance Festival, Easter
- Lord Howe Music Festival, October
- Maitland Hunter Valley Steamfest (steam trains), April
- Moama Village Craft Market, fourth Sunday of every month and long weekends
- Moree Carnival of the Golden Grain, early November
- Orange Machinery Field Days
- Parkes Festival of Sports, Easter
- Port Stephens Festival of the Whales, May to June
- Scone Bushmans Carnival, November
- Sydney Festival, January
- Sydney International Boat Show (Darling Harbour), July
- Tamworth Country Music Festival, January
- Ulladulla Blessing of the Fleet, Easter
- Yamba Family Fishing Festival, September–October

Chapter 2
Deciding What To Take

Decisions made prior to undertaking any form of travel will have a huge impact on the success or otherwise of your holiday. When making decisions, you'll need to keep in mind the purpose of your travel, the type of accommodation you'll be using, the supply of water and energy, your day-to-day living requirements, the disposal of waste, whether or not you're travelling with a pet, and how to keep safe and healthy. All of these aspects require thought, discussion between members of your group, and wise planning decisions – beginning with what to take.

CHOOSING YOUR VEHICLE

Choosing your vehicle is a matter for serious consideration. The choice is complicated by the huge variety available in the marketplace. Possibilities range from small cars (equipped with sleeping swags or tents), to four-wheel drives for towing caravans or trailers, to campervans, to motorhomes fitted out like luxury apartments. From a new vehicle to a second-hand vehicle, to no vehicle at all.

Green vehicle ratings

The Green Vehicle Guide provides ratings on the environmental performance of any new vehicle sold within Australia. This is a website put together by the Federal Department of Transport and Regional Services and the Australian Greenhouse Office that helps prospective buyers choose the most environmentally friendly vehicle suitable for their needs.

The Green Vehicle Guide provides information about environmental performance in regard to greenhouse and air pollution emissions. The ratings are based on government-recognised test standards for measuring and regulating motor vehicle exhaust emissions. Greenhouse ratings are based on carbon dioxide emissions, while air pollution ratings are based primarily on the emission standard to which the vehicle has been certified for the Australian market. These two ratings are then combined to produce a vehicle's final star rating. For more details visit the website <www.greenvehicleguide.gov.au>.

Four-wheel drives

The dream of getting away from it all usually goes hand-in-hand with visions of a shimmering horizon and you at the wheel of a large power-charged four-wheel drive. However, think carefully before you choose this option. Designed for the bush, these vehicles guzzle fuel and emit up to twice the amount of greenhouse gases as do standard passenger vehicles. Their safety record is poor for the occupants, especially regarding roll-overs, and their weight and size means that in a collision they are more dangerous to other road users than standard cars. In addition, loading the roof-rack with water, fuel and other heavy objects heightens the centre of gravity, making roll-over even more likely.

A four-wheel drive's bullbar (also known as a roo bar) is designed to minimise damage to the vehicle at the expense of the kangaroo. In built-up areas, a bullbar will minimise damage to the vehicle at the expense of pedestrians, cyclists and other motorists. Therefore, unless you really do intend spending a large percentage of your time off-road – fording rivers, crossing deserts and salt pans, and scaling sand dunes – consider purchasing a more economical and safer vehicle. Nowadays, 'bush bashing' and tearing up and down sand dunes is seen as environmental vandalism.

If a four-wheel drive is best suited to your needs, joining a four-wheel drive club is an excellent way to learn safe driving techniques, as well as your vehicle's capabilities and limitations. Because these vehicles give access to many pristine places – out of the reach of two-wheel drives – the drivers of four-wheel drives have a special responsibility. The rapidly increasing number of four-wheel drives currently exploring outback areas also means that their drivers must take extra care to travel in a sensitive manner. It only takes a few inconsiderate drivers to spoil the reputation of the majority.

Other travel options

Bicycles, motorcycles, horses and pack horses, horse-drawn 'gypsy' wagons, walking, camels, even llamas – all these and more offer alternative modes of travel to that of motor vehicles. They offer a perfect way to leave the world behind and a more intimate experience of many unique environments.

If due to age, lack of confidence or inability to deal with a breakdown you decide that going-it-alone is not for you, self-drive outback tours are also available. Experienced guides will lead you through the tour with full support regarding planning, four-wheel-drive tuition, emergency medical help and mechanical assistance.

From Robin's travels

On a remote stretch of the Sturt Highway, between Culleraine and Yamba, I spotted a solitary figure ahead. As we drew closer, long grey hair and a beard became visible, along with a wide-brimmed felt hat. An iridescent green safety vest was worn over a heavy khaki overcoat. He pushed a pram-like contraption with a blackened billy hanging on its side, suggesting he led a hard life, a life on the road without modern-day comforts, let alone luxuries. Bounding ahead ran a kelpie, a black and tan dog well able to survive extreme circumstances, like his master.

For the next hour or so, my curiosity knew no bounds. What had motivated the swagman to take up a life such as this? What would it be like to be alone and on the road, with only a dog as company? On our return from Renmark – many hours later and along the same highway – I was surprised to see my mystery man once more. Now at the fruit fly inspection point at Yamba, well into South Australia, he'd made excellent progress, a reminder of the tortoise and hare fable: slow and steady wins the race.

I sensed pride and dignity in his gait, and also fierce independence of spirit. Contentment too. And happiness, maybe. Slowing down as we drove past, we offered a friendly smile and wave of recognition. With a surge of pleasure I received his answering smile and wave, and then he was gone, a retreating form in the rear-vision mirror. This was true green travel, in its most basic form.

PREPARING YOUR VEHICLE

The preparation of your vehicle prior to departure is crucial. It should be checked thoroughly by a qualified mechanic. Any caravan or trailer needs to be checked as well, especially in relation to its tyres, wheel nuts and bearings, brakes, trailer coupling, lights and general condition. A fire extinguisher must be fitted, water containers filled and gas bottles filled and fixed securely.

If your holiday plans involve a small boat, it can be carried on the roof of your vehicle. This will, of course, add more height, wind resistance and load. As far as wind resistance is concerned, though, pointing the boat towards the front will to a degree streamline your vehicle.

Reducing greenhouse gases

The average motorist can reduce greenhouse gases by around 400 kilograms a year by being just 10 percent more fuel-efficient. Some of the ways you can achieve this include:

- keeping tyre pressure on the high side
- using the airconditioner only when needed
- driving less by walking or cycling wherever possible
- choosing your vehicle and trailer or caravan with minimum wind resistance in mind

- avoiding a roof-rack if possible
- keeping your load as low as possible
- keeping your vehicle well maintained
- accelerating slowly
- flowing with the traffic
- avoiding excessive speeds
- avoiding idling your engine unnecessarily.

Basic car repair kit

For general travelling, the following items should make up your car repair kit: can of pressure-pack spray lubricant (for example, WD 40), cleaning rag, fire extinguisher, jack and tools for changing tyres, heavy-duty battery jumper leads, light globes and fuses, radiator hoses, vehicle service and repair manual, set of spanners and screwdrivers, spare fanbelts, spare set of keys (held in a secure position beneath vehicle and able to be accessed without tools), torch with batteries.

Emergency car repair kit

A more extensive repair kit is required if you are travelling in remote areas. Even if you don't possess the mechanical expertise to make the repairs, a kit of this type should still be carried. If you break down, someone else will either guide you through a repair job or do it for you. They won't be amused if you're unable to provide them with the essential tools and materials.

In addition to the basic car repair items listed above, the emergency car repair kit should contain: assorted wooden blocks (for jacking vehicle), brake fluid, breakdown warning reflector, centre punch, cold chisel, cooling-system leak sealer, cordless drill and drill bits, electrical tools (including cables), emery paper, insulating tape and multimeter, epoxy (two part), extra spare wheel, feeler gauges, files (flat, half round, rattail), filter and hose, fuel pump, funnel and gasket material and cement, gauze strainer, grease and grease gun, hacksaw and blades, hammer (medium), heavy towrope and shackles, hose clips, oil (engine, gearbox, differential), oil filter, plastic sheet (two metres by one metre), pliers (regular), water bucket, pressure gauge, radiator insect screen, rope, shovel and bushman's saw, screwdrivers (regular and Philips head), small, medium and large spanners (adjustable), open-ended ring and socket spanners, spark plugs and leads, tyre pump or compressor, vice grips.

PLANNING AND PACKING

Careful planning and packing prior to any journey is an excellent investment of time. Making a list of your needs is a necessity. The following section includes general lists for you to use as a guide for your packing. Personal preferences, the places you intend to visit and the prevailing weather conditions – in relation to the place and the time of year – are, of course, variable, and so too is the available space in which to pack your belongings. Obviously there's a huge difference between packing a backpack to go hiking in the mountains of south-west Tasmania and equipping a motorhome for a journey around Australia taking a year or more.

General tips
When packing, try to place items that are most likely to be needed in easy-to-access places. If remembering where you've put things is a problem, a master plan can be useful; this is especially

so when more than one person is travelling. No loose items should be left that could be thrown around while in transit. This is particularly important in the event of an accident or emergency braking.

When travelling over rough country, a large net can be secured over your load to stop things moving around. When pulling a caravan, loose items can be carried on beds if they are secured with a net. This net can also be useful for protecting gear overnight in windy conditions.

Cardboard boxes – light, flexible and replaceable – are useful when packing items into a cupboard. Wire baskets (instead of shelves and solid drawers) are also an effective way of storing all manner of things.

High cupboards should be packed with light articles only. Make sure that heavy items are stored as close to the floor and the front of the vehicle as possible. The roof-rack should not be packed with heavy items and should be organised to create as little wind resistance as possible.

Keep your packing light and to a minimum, while at the same time be prepared for any eventuality. Overloading your vehicle or caravan is asking for trouble. Before setting out on extended travels a trial run is a good idea. A few long weekends will help iron out a number of potential problems, as well as giving you a good idea as to what equipment is or is not necessary.

When travelling in a group, you can share emergency equipment, water, fuel and rations between the vehicles in your convoy.

Fire safety equipment

Within your set-up, whether it be caravan, motorhome, campervan or tent:

- Install dry fire extinguishers (one by the exit, the other in the sleeping area).
- Place a fire blanket in the kitchen area.
- Install a smoke alarm near the bed.

Emergency food, water and fuel

With sensible foresight and thorough preparation, your travelling will be an exciting adventure. Emergency food, water and fuel are for use only if you're unexpectedly held up due to things such as floodwaters or vehicle breakdown. It's wise to carry an assortment of canned meats, vegetables and fruits, none of which require water to be added. Choose cooked foods, ready to eat.

In order to keep food cool without refrigeration or ice, wrap each item in several sheets of newspaper and store in a cool, shady place.

In remote areas it's vital to carry enough water to last at least five days. Allow four litres per person per day. Water (for both emergency and day-to-day use) should be carried in five- to 20-litre plastic water containers, as well as in built-in water tanks. An emergency solar still can be constructed that will create about two litres of water a day. For instructions on its construction, please turn to page 174.

Extra fuel must always be carried when travelling through remote country. However, do not store any kind of fuel inside your vehicle. Escaping vapours could cause the occupants to feel unwell, as well as creating the potential of an explosion.

From Robin's travels

Using water wisely, as well as carrying an emergency supply, is a basic rule when travelling. On a couple of occasions we've been asked for water by travellers who'd used up their meagre supply. Although we always obliged, we did so with a degree of irritation and reluctance. Carelessness and the Australian outback don't sit comfortably together.

First aid kit

A first aid kit is essential. Commercially prepared kits are available from pharmacies and camping equipment shops. However, making up your own basic kit enables you to select items tailored to your individual needs. A St John Ambulance or Red Cross first aid course will always be reassuring when travelling in remote country, as will a course in cardiopulmonary resuscitation (CPR).

A list of emergency numbers needs to be kept in your first aid kit, as well as in the vehicle's glove box. See pages 156–7 for a list of numbers.

The following list will guide you towards assembling your own kit: adhesive tape (hypo-allergenic) and strips, alcohol swabs, aloe vera gel, antihistamine tablets, antiseptic soap and cream, aspirin, bandages (including triangular), bandaids, burn cream (for use after iced water), constipation tablets, cotton buds, cotton wool, cranberry tablets, current first aid handbook, diarrhoea tablets, disposable hand towels, emergency foil blanket, eye bath, eye drops (single use), eye pads (sterile), forceps, gauze pads, gloves (disposable), hot-water bottle, icepack, insect-bite cream, insect repellent, laxative, manuka honey labelled 'Active UMF Manuka Honey', needle, oil of cloves, paracetamol tablets or capsules (of differing strengths), plastic bags, safety pins, salt tablets, small scissors, sunburn lotion, sunscreen (30 plus), thermometer, tissues, tweezers, vitamin C, water purification tablets, wound closure strips.

When in marine stinger territory, add a two-litre bottle of vinegar to your emergency kit – especially when travelling between the months of October to May.

Pack your first aid kit within an unbreakable plastic box with a strong, well-fitted lid. Place within your vehicle, in an easy-to-reach position.

Lifesaving device
A large plastic bottle (containing a little water) attached firmly to a reasonably long length of rope can be a lifesaver. Place it in your vehicle in an easy-to-reach place. If the need arises it can be thrown into water to assist someone in trouble. Likewise, it can be used to rescue someone stuck in quicksand.

Important documents and equipment
ATM cards, cash, cheque book, credit cards (check expiry dates), drivers' licences (check expiry dates), glasses (sun and reading), insurance papers and last year's tax return (if travelling for an extended period), Medicare and/or healthcare card, motor vehicle registration papers, notebook and pencil, prescriptions for medications, wallet or purse.

Glove box
Your glove box is perhaps the best place for some of the important documents listed above, namely your vehicle registration papers and insurance policies. The following items are also useful:

compass, magnifying glass (for lighting a fire), maps, mirror (for emergency signal), notebook, pen and pencil, torch, waterproof matches.

Basic foods
Biscuits, bread, butter or margarine, cheese, coffee, cooking oil, cold drinks, flour, fresh fruit and vegetables, fruit (tinned and dried), jam, meat, milk (UHV), oatmeal, pasta, rice, salt and pepper, soup (tinned and dried), sugar, tea, dried herbs and spices, Vegemite.

General camping gear
Awning (with poles, pegs and ropes), backpacks, binoculars, camera (including lenses and film), camp shower and toilet, chairs, clothes line and pegs, doormat, electric radiator, extension leads, fire extinguishers, fly nets, funnels, mattresses, mobile phone, radio, shovel, table, tarpaulin (plastic), tomahawk, torch (with spare globes and batteries), trowel, water hose.

Clothing
Gloves, warm hat, jeans, long-sleeved shirts, parka, scarf, shorts, skivvies, slacks, socks, sunhat, sweaters (woollen or polar fleece), swimming wear, thermal or woollen underwear, thongs or sandals, T-shirts, cotton underwear, walking shoes, waterproof jacket and pants.

Nightwear and bedding
Blankets, doona or sleeping bag and swag, sheets and pillowcases, sleepwear, slip-on shoes, extra woollen sweater and warm socks for extremely cold nights. (You may decide to sleep under the stars, in a modern swag fitted out like a tiny tent, with rain covers, an inbuilt mattress and mosquito screens.)

From Robin's travels

While packing for a September journey to the Flinders Ranges, I asked Doug what bedding he'd like me to pack. 'What are you taking?' he asked. I told him. 'Well, I certainly don't want a doona,' he said. 'A couple of blankets will be plenty.'

On the second night there was a heavy frost. After living on Phillip Island for five years, we were unused to this type of temperature extreme, and Doug began to ask, somewhat anxiously, about spare bedding. There was good news and bad. I'd packed extra blankets for the dogs, but not for Doug. For the remainder of our wanderings Doug's bed was made amply warm with the addition of heavy, dark-grey ex-army blankets. Clean, of course.

Sewing kit

A small collection of different-sized buttons, needles and threads (cotton, synthetic and wool), pins, safety pins, scissors, tape measure and zips.

General kitchen items

Bottle opener, can opener, coffee plunger, chopping boards, colander, corkscrew, cutlery, cutting knives, funnel, glasses, mugs, plates (large, small and bowls), spatula, spoons (serving and wooden), tea towels, thermos, trays (for eating meals off), water filter or jug.

Cooking gear

Electric, gas and white spirit appliances

Double-adaptor plugs, electric frying pan, electric jug, fuel cooking stove and oven, frying pan, fuel (gas bottles or white spirits), matches, microwave oven, paper towel, plastic bags, pot holders, saucepans (small, medium and large).

Campfire

Store the following items together in one place, preferably so that you can access them from the outside of your set-up: barbecue grid and plate, barbecue tongs, camp oven, firewood and kindling, frying pan, kettle, large dixie for heating water, long-handled toasting fork, nest of billies, newspaper, paper towel, plastic bags, sturdy pot holders, waterproof matches.

Green cleaning and pest control products

Bicarb soda, borax, bucket, cloves, dishcloths, eucalyptus oil, flyswats, full-size soft broom, garbage bags, hand broom and pan, laundry and kitchen detergents (biodegradable, phosphate-free), scourer, pure soap, sponges, steel wool, vinegar (white), washing-up basin, washing bucket with clip-on lid (for washing clothes), washing soda (sodium carbonate, sold as 'Lectric' soda).

Personal items

Brush and comb, hot-water bottle, insect repellent, mirror, moisturiser, nail file and scissors, prescription medicines and authorities, razor, reading glasses, safety pins, shampoo, soap (in container), sunscreen (30 plus), sunglasses, tissues, toilet paper, toothbrushes and paste, towels (full size, hand and face washer), watch or clock.

Note

When purchasing paper products (tissues, toilet paper and kitchen paper) make sure that the label reads: recycled paper, not rebleached, with 100 percent biodegradable, recyclable packaging made from 100 percent recycled material.

Recreational requirements

Address and telephone book (friends, relatives,

repairers), bird book, other books and manuals
for leisure and information, calendar or diary,
hobby gear, magazines, maps, medical reference
books, paper, pencils and sharpener, pens, permits for camping, permits for entering land
owned by Aboriginal communities, fishing equipment, fossicking equipment, snorkelling
equipment, water bottles for daytime use.

The well-known saying 'A place for everything
and everything in its place' makes a lot of sense –
especially while travelling, when space is limited.

With your vehicle chosen, checked and serviced,
your possessions packed and your route mapped
out (taking into account your health, age and
experience), safe, stimulating travel is an achiev-
able goal.

SHOPPING ALONG THE WAY

There are of course some things that can't be
packed, such as a constant supply of perishable
food. However, it's always rewarding to shop in
small country towns. The local butcher will be a
wealth of information and opinion, especially if
you begin the encounter with genuine praise for
his town and area. Soon he'll be slicing you his
finest porterhouse steaks and you'll leave with a
smile on your face and a real 'feel' for the town –
along with no plastic or polystyrene packaging.

Even better in terms of greenness is the barter
system, especially when it applies to staple food
items.

From Robin's travels

At a butcher's shop in Renmark, South Australia, we
observed a man enter with a cardboard box and

exchange two cabbages (so fresh the sap was still running) and a bunch of parsley for eight sausages. And all with a smile!

When we asked this same butcher for some pork chops, T-bone steaks and two lamb shanks, he hesitated then inquired, 'What do you want the shanks for?'

Surprised, I replied, 'A stew. We love the flavour.'

'That's all right then,' responded the butcher. 'I haven't got many and was hoping they weren't for your dog.'

After chatting about ingredients and the process of cooking in a camp oven over a bed of hot coals, he reached for a couple of sheets of white paper, jotted down the amount for each purchase, added up the total, subtracted a little, and then wrapped the parcel. 'Happy travelling,' he said, 'and I hope you enjoy the shanks.'

Pride in their produce is also a feature of many small town fruiterers. Enter their shop and you'll be treated to the aroma of fresh, flavoursome produce. Praise their produce and you'll be directed towards the choicest fruit and vegetables, along with locally produced honey, homemade jams and chutneys. With your calico bags full, you won't have needed any plastic packaging at all.

Honesty systems

Some roadside locations offer travellers fresh fruit, vegetables and flowers with an honesty system of payment. Many small camping areas operate a similar honesty system whereby travellers make their payments into a sealed box. It's a measure of Australian pride that the large majority of tourists honour this system. Stinginess is not Australian.

Bulk buying

Whenever you come across a health food shop that sells in bulk, it's a wise green approach to stock up on all your supplies. When you leave – with your purchases contained within your own cloth bags – you'll feel pleasure that you're doing your bit towards reducing the use of unnecessary packaging.

Chapter 3
Selecting Where To Stay

The Australian outback can be both magnificent and disturbing at the same time. While some people find the space liberating, others feel anxious and find themselves heading for the nearest town. In recognition of this difference, the tourism industry offers accommodation of all types: from caravan parks and accommodation in population centres, to exclusive tree houses and the solitude of bush camping.

The number of options available for the overnight accommodation of travellers is almost endless, from the most basic to the most sophisticated. You can choose from swags beneath the stars, tents, camper trailers, conventional or pop-top caravans, campervans, motorhomes and even horse-drawn gypsy wagons. Houseboats, cabins, bed and breakfast accommodation, backpacker lodges, on-site vans, self-contained units, hotels and motels offer another set of accommodation options.

Ecotourists know what type of accommodation they want and are prepared to pay for things to be done correctly. They want their travel to have minimal environmental and cultural impact, while at the same time being economically viable for the operator. Ecotourists don't want to see or smell sewage discharge, stormwater overflow, plastic litter, estuaries clogged with silt or coloured by algal blooms, dead and dying seagrass, skeleton-like forests of red gum, salting or erosion.

ECO-RESORTS

Eco-resorts are shaping up to be the latest trend in accommodation. Staying at an eco-resort doesn't mean you have to pay for uncomfortable, primitive surroundings and do without modern conveniences. On the contrary, these establishments are frequently architect-designed and are usually very comfortable and stylish. In fact, many eco-resorts are luxury retreats that offer peace and quiet as an added bonus. Frequently nestled amongst trees, these thoughtfully designed retreats have environmental concerns foremost in mind. Most are located in beautiful areas and treetop balconies are popular. Many offer extra recreational facilities, along with the lure of a natural wilderness environment.

Ecotourism isn't about minimalistic living, but rather an intelligent use of the resources of that particular area. For example, in some localities water can be so abundant that it can be used without constraint. If your preferred eco-accommodation is situated in a place with high rainfall – for instance, Flinders Island or King Island in Bass Strait – water rationing will seldom be required. High rainfall and low population density mean that rainwater tanks, bores, natural springs and mountain streams usually provide abundant water.

The cleanliness of hotel rooms and cabins marks them, in the eyes of most travellers, more than any other single thing. Nail clippings beside the bed, toast crumbs in the cutlery drawer, strands of hair in the wash basin, mould in the shower recess – all these and more are inexcusable. What is not necessary, however, is the heavy use of chemical cleaning products to maintain high standards of cleanliness. The use of biodegradable products and the application of as little product as possible is now accepted practice, along with thorough vacuuming.

Ecologically sensitive accommodation will include a blend of the following features.

- The process of establishing buildings and gardens will have involved minimal clearing of native vegetation and minimal soil disturbance.
- The use of recycled timber and timber products, mud brick or stone will be used in construction wherever possible.
- Termite control using natural barriers rather than residual poisons will be implemented.
- Verandahs and eaves will capture winter sunshine and also give shade protection during summer.
- Building design will ensure good cross-ventilation, plenty of natural light and the use of generous quantities of wall and roof insulation using pure wool or other natural or recycled materials. Eco-ply cladding may be used.
- Wood stoves may be installed in areas where firewood is abundant. These should have solar mass walls.
- The planting of dense natural barriers of vegetation between units may be used to reduce noise.

Native tree and shrub plantings, as well as needing little water, will attract native birds.

- Self-composting toilets or an onsite water treatment plant will filter and treat sewage and all other waste water, with subsequent re-use of this grey water in the garden.
- Solar power will be used throughout, or for hot water and heating. Alternatively, electricity may be provided from an environmentally friendly hydro-generator set up on a nearby creek, or from a wind generator.
- Rainwater will be used for bathing and filtered rainwater or UV-filtered mountain water for drinking. Sinks will be small in size and shower heads low-flow.
- Organic food may be grown on an adjoining biodynamic or permaculture farm, using non-chemical weed eradication. Kitchen waste will be composted for use on fruit and vegetable gardens, incorporating the separation of oils and fats from the waste to prevent soil degradation. Alternatively, a worm farm may be used to recycle kitchen waste.
- Other rubbish will be recycled whenever possible, products will be bought in bulk to reduce packaging, and paper products that come from recycled, unbleached paper waste will be used.
- Pure cotton and pure wool bedding and linen, and soaps and shampoos made from natural products may be provided. Biodegradable, phosphate-free washing powder and cleaning products will be used.
- Bicycles may be provided for use by guests.
- Some establishments may even boast no TVs or phones!

Sources of information

If you are planning to stay in eco-accommmodation, you may want to explore the wide range of options available: from caravan parks and camping grounds, to top-of-the-line luxury accommodation. The following organisations, tours and nature retreats will help you with your research.

Ecotourism Association of Australia

The Ecotourism Association of Australia is the peak national body for ecotourism in this country and offers an eco-certification program for tourism operators in Australia – a world first. Being energy efficient and using the latest water saving technology brings financial rewards for both the operator and tourist. To search for accommodation certified by the association, visit their website at <www.ecotourism.org.au>.

You can also search for other ecotourism products such as nature-based attractions and guided eco-tours on this website.

Nature and Ecotourism Accreditation Program (NEAP)

Tour and accommodation businesses can undergo an accreditation process organised by the Victorian Tourism Operators Association, called the Nature and Ecotourism Accreditation Program (NEAP). There are various categories of excellence. If you'd like more information, check out the website <www.vtoa.asn.au> and follow the links to the accreditation program.

AAA Tourism

AAA Tourism is the national tourism body of the Australian Automobile Association (AAA). This

organisation assesses a wide range of accommodation Australia-wide, including hotels, motels,
lodges, cabins, caravan parks, camping grounds,
apartments, holiday units and five-star resorts, as
well as guided tours. They offer details on more
than 14,000 accommodation options from all over
Australia.

Along with a property's original STAR rating (a
one- to five-star rating depending on the standard
of facilities on offer), a Green STAR endorsement
has now been introduced that rates a facility's
environmental practices. A Green STAR rating is
used to demonstrate the level of environmental
awareness and commitment to sustainability of
an operator.

Sixty-two percent of travellers say a green
endorsement would encourage them to choose
one accommodation over another. Under this rating system, a green property must include
features such as:

• energy efficiency, in relation to lighting, heating
  and cooling, and natural ventilation
• waste minimisation, in relation to recycling
  rubbish, and towel and linen re-use option
• water management, for example the use of dual
  flush toilets, environmentally friendly cleaning
  products, unbleached paper products and recycled water on gardens.

For more information, visit AAA Tourism's website at <www.aaatourism.com.au>.

Youth Hostels Australia (YHA)
This organisation plays an active role in environmental conservation by offering eco-hostels where
travellers can experience sustainable living.
Minimising waste, recycling, reducing noise

pollution, promoting energy efficiency and water conservation and encouraging the use of public transport all play a role in caring for our environment. If you'd like to know more, check out the website <www.yha.com.au/australia/environment.cfm>.

Ecotourism in Tasmania
Tourists visiting Tasmania are in for a treat. A number of operators offer Gordon River wilderness cruises, giving visitors the ultimate eco-tourism experience. A range of accommodation is available in Strahan, the town from which these cruises operate.

For more information visit the websites <www.tasforestrytourism.com.au> and <www.discovertasmania.com>, or phone 1800 800 448.

From Robin's travels

Whenever I read advertisements for guided walks around Cradle Mountain, memories come flooding back of my own Cradle Mountain hike in the January of 1966. There were no private huts with hot showers, gourmet meals or comfortable beds. Instead, we slept in hike tents, or on the rough floorboards of mountain huts. Our meals were made up of dehydrated foodstuffs; we washed in icy cold streams and endured a heat wave of temperatures well over 30 °C, then a snowstorm – all in the space of ten days. The challenges were to remain happy, fit and healthy; to outsmart the rogue brush-tailed possums that raided our packs; to take care of others in our group; and to experience the enchantment of this ancient, alpine wilderness.

Destinations

There are many destinations in Australia for those wanting to make as little impact on the environment as possible while they travel. The following destinations are examples of the achievements that the development of ecotourism has seen.

Shark Bay, Western Australia

Monkey Mia is famous for its bottle-nosed dolphins. What is less well known is that raw sewage from the tourist facility once killed a number of these unique dolphins. Nowadays sewage is treated and piped inland to be absorbed into the vast tracts of red desert sand.

Since Shark Bay achieved World Heritage status, only green cleaning products have been used throughout the extensive complex. It is a modern, environmentally friendly facility, built to cater in a sustainable way to the demands of the large numbers of eco-tourists seeking the dolphin experience.

Here the blood-red desert meets the shimmering blue of the ocean and careful use of water and electricity is essential. Both water and electricity are scarce resources at Shark Bay and expensive to provide. Diesel generators provide power and a desalination plant supplies washing and drinking water. Visiting the dolphins comes at a price.

Lord Howe Island

Only a two-hour flight from Sydney, Lord Howe Island gives travellers a unique experience in a magnificent environment. With its World Heritage listing, mild climate, lush tropical rainforest (made up of palms, ferns, mosses, orchids, dramatic banyan trees and pandanus), coral reefs and unique woodhens, Lord Howe Island offers accommodation of all types except camping, which is not permitted anywhere.

Lord Howe Island is one of the top nature-based and ecologically sustainable destinations in the world. With very little traffic – most people either cycle or walk – and only 393 tourists allowed at any one time, a peaceful, litter-free environment is a reality. Environmental initiatives include strict waste recycling; a vertical composting unit; a bulk food cooperative aimed at decreasing packaging and so waste; and in the planning stages, a wind-powered electricity generator.

Limiting the number of tourists to a particular area is a sound means of keeping a fragile environment intact for future generations. As mentioned, Lord Howe Island uses this method as a means of protecting its World Heritage listing status. Rottnest Island also limits its permanent population, as does Norfolk Island.

On Lord Howe Island, a balance has been achieved that protects the island's unique environment while at the same time catering to the needs of the community and eager ecotourists.

Bogong Horseback Adventures, Victoria

At Mount Bogong in Victoria, an enterprising couple offers adventure treks into the high country, using pack and riding horses. Sleeping under the stars can be a feature of this experience, with waterproof swags providing protection from mountain mists or the deep chill of alpine frosts. Campfires, icy mountain water, historic cattle-man's huts, wildflowers and awesome mountain vistas take visitors far beyond themselves.

During the summer months (December to April), this award-winning enterprise offers great food and wine, mountain-bred horses and friendly, experienced staff. Three to seven day tours are on offer, as well as two-hour, half-day and full-day rides. These packhorse-supported tours take you into an alpine landscape inaccessible to four-wheel drives.

Bogong Horseback Adventures depend on a unique yet fragile alpine environment. Routes and itineraries are chosen, therefore, depending on the time of year and weather conditions in relation to the erosion of tracks. Fire risk is kept in mind, and low impact riding and camping practices are encouraged. If you'd like more information, visit the website: <www.bogonghorse.com.au>.

Healing Dreams Retreat
This nature-based retreat is located on Flinders Island, in Bass Strait. Catering for eight to 15 people, this island resort is set on 105 hectares and includes an organically certified orchard and vegetable garden, rainforest, streams and waterfalls. Perched on the side of Mt Strzelecki, the resort has stunning ocean views, peace, tranquillity and some of the cleanest air and water on earth. Healing Dreams Retreat is known as one of the best wilderness retreats in Australia – along with providing luxury accommodation, a gym, outdoor spa and many other facilities.

Caring for the environment is foremost in the minds of the owners. Towels and bed linen are 100 percent cotton and their natural soaps and shampoos are made on the island. The only noises you'll hear are the sounds of water and the songs of birds. Rainwater is collected then purified within a large ultraviolet system. If you'd like to know more, check out the website <www.healingdreams.com.au>.

Promway Horse-drawn Gypsy Wagons

An enterprising family business in Yarram, Victoria offers horse-drawn gypsy wagons for hire, pulled by placid Clydesdales. This home-away-from-home gives people the experience of living like the pioneers. Solar bush showers add to the green nature of this holiday with a difference. Well-behaved dogs are welcome.

CAMPING IN CARAVAN PARKS

Planning the trip

Depending on the personality types of the people doing the travelling, the degree of planning required will vary. Too much planning can take the adventure out of travel, but for some people uncertainty can lead to anxiety. Most people are familiar with the stress brought on by being unable to find a suitable place to spend the night. It's not a good feeling and can spoil what should be a relaxing holiday. For this reason, many travellers book ahead so that every single night is covered. There are many books that list caravan parks Australia-wide and also provide information about their facilities. You can plan an entire journey from the comfort of your lounge room, using these books and your telephone or computer. However, by doing so you commit yourself to an itinerary that cannot be changed.

Some travellers set themselves busy travel schedules, leading to an exhausting itinerary. What if someone feels unwell or needs to rest? What if you love the area and want to stay longer? Seeing everything and going everywhere is not necessarily a recipe for successful travel – travel overload is a common result of this type of itinerary. It's usually more satisfying to experience the true feel of a place and its people. By selecting your campsite carefully and then spending time getting to know a place, you allow rich observations and friendships to develop.

If you prefer less structure to your travel, you can choose instead to plan only one or two days ahead. Using books that list caravan parks gives you the opportunity to inspect a number of parks in a particular locality and then take your pick. A flexible itinerary allows you to stay longer in favourite places and act on recommendations offered by fellow travellers.

Age can be a factor when it comes to deciding how much of a trip to plan, as can health. If this is the case, you will probably need to plan more carefully, to ensure reasonably close proximity to medical services.

Standards

Caravan parks vary enormously in the quality of their management, amenity blocks and general upkeep of gardens and grassed areas. Those parks with a lower percentage of permanent vans generally have superior facilities for travellers and a more relaxed holiday atmosphere. Those listed in the most popular books usually pass a certain standard. Some books come with a star rating. See page 38 for more details about STAR ratings.

Socialising

When selecting your campsite, choose a place that allows other campers their privacy. Nothing's worse than having noisy campers arrive after dark and set up camp right next to you – especially when plenty of other places are available. If you see someone camped alone, don't assume they'd enjoy your company. The fact they're camped alone usually means that's how they prefer it.

On the other hand, an opportunity to socialise with fellow travellers is on offer at most caravan parks across Australia. Some lasting friendships will develop, as will many brief but pleasant encounters along the way. Exchanging travel experiences, swapping information about places, weather conditions and people, discussing vehicles and towing options, sitting in the sun and talking about life – the possibilities are endless. A melting pot of careers, lifestyles and expectations are to be found within the parameters of most caravan parks. A large percentage of travellers require the facilities on offer at caravan parks and value the security of fellow travellers.

Spending more than a night in one place increases your chances of meeting interesting people. Every person has a story. Each new traveller adds richness to the tapestry of your life. It doesn't matter that you won't meet again. The important thing is, you did meet. You made a connection.

From Robin's travels

Rawnsley Park Station, located in the Flinders Ranges, South Australia, offers a range of accommodation, from cabins and onsite caravans to a caravan park with power, as well as an extensive area for bush camping. There is a swimming pool and a general store that also sells fuel. This station has changed from farming sheep to 'farming' tourists. As a result, the property is now sustainable. The shearing shed has been converted into a restaurant, whilst maintaining a small section for shearing demonstrations.

BUSH CAMPING

Camping in rest areas, state forests and council reserves

Each state of Australia has different laws regarding overnight camping in rest areas. These laws are constantly changing. Consequently it's wise to observe signage and respect people in authority, while at the same time using common sense.

A good range of facilities such as toilets, tables, seats, shelters and fireplaces or barbecues are found at many roadside rest areas, roadside camping areas, state forests and council reserves. Shelter and shade trees are usually available. Rubbish bins may or may not be provided. Some provide fresh water. A small fee may be charged, especially if a shower is provided. Rules and regulations control these areas and must be respected. Camping permits are required, in some places. Some parks are for camping only, and do not allow caravans.

Camping in national parks

National parks offer the privilege of camping in some of the most beautiful and environmentally precious sites throughout Australia. Rules apply and must be strictly obeyed:

- Camp only in designated areas and at least 20 metres from any waterway.
- Keep to walking tracks to reduce tramping of vegetation.
- Do not drive vehicles, including motorcycles, off formed roads.
- Do not disturb or remove any plants, animals, archaeological sites and geographic features. These are protected by law.

- Do not bring any firearms, generators and chainsaws into the park.
- Leave the site clean and tidy.

For many national parks, fees apply and permits may be required. Some parks are for camping only, and do not allow caravans.

For further information, visit:

- the Australian Tourism Network at <www.atn.com.au/parks/index.htm>. This website has comprehensive information on national parks in all states, with links to other sites.
- the Australian Government's Department of Environment and Heritage at <www.deh.gov.au/parks>. This website includes information on national parks, marine protected areas and botanic gardens throughout Australia.

Protecting wildlife and vegetation

The use of signs, boardwalks, walking trails, protective barriers and educational boards guide travellers towards awareness of the environment. They also protect wildlife and flora from introduced animals and plants – and tourists.

Never feed or tease native animals. Although seemingly tame and friendly, native animals can be unpredictable and aggressive. Human food is never appropriate for wild animals, and hand-feeding teaches native animals to trust people. This frequently leads to tragedy.

From Robin's travels

The Australian environment is constantly under threat from introduced plants and animals, as well as people. While camping on the banks of the Murray River,

I watched as a feral goat nimbly descended a steep
embankment to drink of the Murray's lifeblood.
Smelt the rank odour of a fox. Saw a heap of carp
carcasses rotting on the bank. Recognised the distinc-
tive pawprint of a cat, stamped on a sandbar. Heard
the rhythmic beat of a pump redirecting the Murray's
water to a dried-up billabong. A billabong dense with
thistles and surrounded by hundreds of stressed,
dying and dead gums. Then, in contrast, after soft rain
I heard the thump of kangaroos and recognised hundreds of thousands of minute red-gum seedlings
beneath my feet and carpeting the ground.

Free bush campsites

Campsites that cost nothing are scattered
throughout Australia, and more and more travellers are choosing to camp in these places. There
are books and booklets listing these campsites
and most include maps. These books also contain
information regarding availability or otherwise of
toilets, water, fireplaces, tables and seating, as
well as their dog-friendly status. You can purchase these books in camping stores, book shops
or by mail order. They are included in the Further
Reading section on page 221.

Closely guarded secrets

There's a tendency to keep quiet whenever you
discover the perfect bush camp. Your motivation
may be a passion to preserve the fragile environment of that special place; or alternatively, a
selfish desire not to share. Sometimes it's difficult
to tell the difference! When asking locals about
bush camping sites, though, it's easy to recognise
the glazed look that usually means they're not
going to disclose the information.

Saving money

Saving money is perhaps the most common motivation for people using free camping sites, but there are other reasons too. The increasing use of these places may be linked to the relative popularity of porta-potties, solar energy, hot-water shower systems (solar and gas) and small lightweight generators. With these systems, it's easier for people to go bush, while at the same time camping with many home comforts. In addition, the number of people travelling with dogs has increased, as many people believe that the presence of a dog makes bush camping a much safer option. Surveys show that most long-term travellers free-camp at least half of their time away.

Bush camping issues

There are many things to consider when camping in the bush. A group of less than eight people will have a reduced impact on the environment, as will camping off-peak. The accidental spreading of disease and noxious plants via vehicle tyres, muddy boots and tent pegs (for example, rootrot fungus), or socks (for instance, burrs of all types) must be kept in mind.

When you leave a campsite, it's easy to leave it tidy. It's more of a challenge to leave no trace at all. Perhaps we all need to feel we're the guardians of these free camping places.

When bush camping remember to:

- Obtain permission prior to camping on private property, including Aboriginal land.
- Look for campsites on foot in places where road conditions are unknown or appear doubtful

(such as mud or deep sand), especially if towing a caravan with a two-wheel-drive car.
- Choose your site two to three hours before sunset.
- Choose a level, hard surface for your campsite so you'll leave no imprint. Old roads, gravel pits and dry saltpans can be ideal sites that are also safe for fires.
- Avoid driving off-road vehicles and motorbikes unnecessarily along watercourses or over virgin soil. Avoid compacting or eroding the soil and crushing plants and animals.
- Understand and respect local weather conditions.
- Choose a designated camping area, an existing campsite or a natural clearing rather than create a new one. Permanent structures are never allowed.
- Camp at least 20 metres from any river, creek, billabong, reservoir or lake's edge.
- Camp on the sheltered side of a hill or sand dune to help protect you from wind and dust.
- Choose a campsite with a northerly aspect during the months of winter, when sunshine is a bonus. During summer, a southerly aspect is more appropriate.
- Always obey signs, especially if they warn you about crocodiles. Invading a saltwater crocodile's territory can be dangerous. Other creatures to be cautious about include European wasps, feral pigs, dingoes and camels. Their footprints, scratchings and droppings will alert you to their presence. See Chapter 11 for more details.
- Avoid disturbing sheep, cattle and wildlife while they drink, so do not camp near stock

watering places such as tanks, dams, bores, troughs or windmills. These locations are also problematic due to the proliferation of mosquitoes and flies – these insects are attracted by water and high concentrations of animal droppings.

- Avoid dense tea-tree and mangroves as they frequently harbour mosquitoes, sandflies and midges.
- Take a careful look at the ground, checking for ant mounds and ant tracks.
- Position your caravan doorway on a piece of slightly raised, well-drained ground, with the door facing away from the prevailing wind.
- Never set up camp in a dry creek bed. If heavy rain occurs in the upper catchment, torrents of floodwater, with walls up to one metre high, can sweep down a dry creek bed without warning. Flash floods kill.
- Always consider the health and age group of your participants when planning to go bush camping. Getting away from it all is always a refreshing experience. Remoteness can, however, be a safety issue if things go wrong.
- Make sure your caravan is level, as this will affect the quality of your sleep. It's wise to carry with you some pieces of timber that you can place under the wheel when you jack up the lower side of caravan.

Native vegetation, birds and animals
Native plants and animals are protected by law. Never cut down or damage live trees or vegetation, or disturb the earth unnecessarily. Modern tents (with their sewn-in floors) seldom need trenches, especially if you choose well-drained or raised sites. Before pitching your tent, remove

loose vegetation and rocks from the site and put them carefully to one side. When leaving, replace the vegetation and rocks so the site is restored as much as possible to its original state.

Protecting native vegetation, birds and animals must always be a priority. Feeding or attempting to tame native animals and birds is never kind. The dingoes of Fraser Island demonstrate the danger of this practice.

From Robin's travels

Stumpy-tailed lizards, affectionately known by locals as sleepy lizards, are very much part of outback travel. These docile reptiles have developed the dangerous habit of sunning themselves on outback roads. We often stop to move a pair of basking lizards to the relative safety of roadside vegetation. It's distressing to see so many lizards 'ironed' dead by car tyres. With captive-bred blue-tongued and stumpy-tailed lizards at home, our dogs accept the wild stumpy-tails as simply part of the scenery. However, we are delighted by the lizards' lazy charm, their socialising in distinct pairs and their opportunistic scavenging around campsites.

River red gums

Sometimes called 'widow makers', these trees are infamous for dropping entire branches without warning, even in dry still weather. River red gums need to be treated with great respect. Trees with dead or hollow branches or trunks are especially dangerous. Although trees can lose weakened limbs at any time, strong gusty winds make this more likely. Therefore, look upwards before choosing your campsite.

From Robin's travels

Australia's most important inland waterway, the Murray River, offers approximately 2500 kilometres of river to explore. Forming a natural border between Victoria and New South Wales, many important towns have established themselves along its banks. Although magnificent river red gums are a feature of this mighty river, signs warn campers to avoid setting up camp beneath its branches. I admit I've been careless at times, but never again. Strolling along the riverbank while enjoying the golden hush and stillness of sunset, a huge branch fell with a mighty splintering crash only metres from my retreating heels. I will always look above me now, especially when selecting a place to camp.

Safety

On Friday and Saturday nights in particular, it's a good idea to avoid bush camping within a 30-kilometre radius of any large town. This will reduce the likelihood of being disturbed by a carload of hoons out for a night of heavy drinking. When preparing for the night, lock your vehicle and then, if you're sleeping inside a campervan, caravan or motorhome, lock the door, leaving the key in the inside of the lock. If worried about your security, don't step outside. If possible, drive off and telephone police. Many people travel with a dog as it gives them a feeling of added security.

Leaving a campsite (van or motorhome)

Before leaving your campsite there are a number of things to check. These include:

• The load is correctly balanced.

- All loose items on sinks and shelves are put away.
- Cupboard doors and drawers are secured.
- All liquids, including those in the refrigerator, have secure lids.
- The refrigerator door is locked.
- The gas bottle is turned off.
- The roof hatch is closed and locked.
- All windows and flyscreens are closed and locked.
- The jockey wheel is either secured or removed.
- Jacks are raised or safety stands are removed.
- Power lead, water supply hose and drain hose are all disconnected.
- The cap on the water tank is secured and locked.
- All lights are tested and working properly.
- Safety chains are connected.
- Wheel chocks are removed.
- Tyres, wheels and brakes are tested and in good working order.
- Door footstep is lifted and the security hook connected from door to van.
- External door key placed in the towing vehicle.
- Finally, your campsite is left in better order than you found it.

Idling your car engine
Some people have the annoying habit of idling their car's engine (for 15 minutes or so), prior to departure. This practice is inconsiderate to fellow travellers and the environment – car exhaust fumes are toxic. The sound of a car's engine, unnecessarily idling, drowns out birdsong and shatters the peace of the bush.

Chapter 4
Clean Water

The human body can survive for many weeks without food; however, without water, most of us would die within a week or so. It's generally recommended that an average adult in a temperate climate needs to drink about 1.5 litres of water a day. Some water is, however, so polluted that it will kill you rather than provide essential fluid.

Generally speaking, there are three grades of water purity:

1 Water safe for drinking, cleaning teeth, gargling, making ice, washing salad fruits and vegetables, cooking, and washing your face and hands.
2 Water used for washing dishes, washing clothes and showering.
3 Water that's too polluted to use for anything.

Many types of viruses, bacteria, algae and minute parasites contaminate water. Some are microscopic, while others can be seen with the naked eye. Even water sourced from the most pristine wilderness environment may contain micro-organisms capable of causing serious disease, ranging from mild to serious gastroenteritis (diarrhoea and vomiting), to potentially life-threatening diseases such as leptospirosis, dysentery, cholera and hepatitis.

As travellers we need to be forever mindful of water, both in relation to its safety and scarcity. Travellers cannot assume that all water is safe to drink, or even safe to use for showering. Whether our water is bottled, town, rain, bore, well, river or lake it's a resource that must be guarded from contamination.

Unsuspecting travellers, campers and hikers sometimes drink water containing the cysts of parasites such as cryptosporidium and giardia. Shortly afterwards, these waterborne protozoan parasites 'set up camp', causing symptoms such as watery diarrhoea, abdominal pain, stomach cramps, vomiting and fever – lasting from three days to six weeks or longer. This type of gastroenteritis can cause serious problems for people with either an existing health problem, or an immune system already weakened by disease. Young children, seniors and those with AIDS or transplants, also need to be especially vigilant.

WATER SOURCES

Town water

For most of us – living in settled areas throughout Australia – safe town water is something we take for granted. Most town-water supplies have careful catchment management and well-engineered treatment plants that add chemicals such as chlorine to safeguard the supply against harmful organisms such as giardia and cryptosporidium. Most people are familiar with the cryptosporidium and giardia outbreaks that occurred in Sydney from July to September 1998.

As well as adding chlorine, fluoride (which protects children's teeth from decay) is added to some town-water supplies. Town supplies may also be treated to improve the water's colour, taste and odour. For these reasons, many people prefer to drink bottled water in spite of its cost. Bottled water usually has a better taste than tap water, but it pays to check the label as some bottled water is simply filtered town water.

Lake, river, creek, well, bore or dam water

When water from these sources has been directly or indirectly contaminated with human faeces (as a result of septic tank seepage and waste) or animal faeces (run-off from pastures or feedlots, or bird droppings), bacterial contamination of the water with microscopic single-celled organisms such as E. coli occurs, as well as the risk of contamination by giardia and cryptosporidium. Therefore, this water is seldom safe to drink.

Murray River water is considered suitable for cooking and drinking, provided it's been treated. This river system is in dire need of care. As a camper along its banks, you must ensure that no human or pet faeces, litter, soap or detergent pollutes the river or its tributaries. When bush camping, use biodegradable cleaning products, and use minimum quantities. The waste water is best poured into a shallow pit dug into the soil, at least 100 metres from the river. In order to absorb grease, it's a good idea to partly fill the pit with dry leaves and twigs. Every few days the leafy matter can be burned in your campfire. Refer to pages 118–19 for further information about the safe disposal of waste water.

The water found in rivers and lakes is usually safe for washing clothes and showering, as well as washing car windscreens and lights. With water such a scarce and precious resource, travellers need to be sensitive in their use of water, and aware of local supply. As a group, tourists have a bad reputation when it comes to their use of water. Locals need to be shown that guests in their area will not waste water, especially when restrictions are in place. Local radio offers insight into water issues in a particular area.

If you're taking water from a lake, collect the deepest, coldest water possible. Cold water contains less micro-organisms than warmer water. To avoid surface scum, collect water from the side of the lake not facing into the wind, and reach out, away from the edge. If you can, collect water from the end of a jetty.

Running water taken from a stream may look cleaner than lake water, but this is rarely the case. Many of us remember the pleasure of drinking from ice-cold streams while hiking in the mountains. In most places this practice is no longer safe. You can cool off, though, by soaking your feet and wrists in the stream.

From Robin's travels

One of our favourite camping places on King Island was located right on the beach, nestled against steep, heavily vegetated sand dunes. With the deep, soft sand as a mattress, we slept under the stars and listened to the trill of penguins tumbling out of the waves. There was sea water to bathe in, abundant driftwood for

campfires, salmon to catch in the surf – and the
world's purest spring water. Trickling through mossy
beds and out onto the sand, this water was filtered
through 250 acres of sand-dune country. It had a
quality all of its own. Nothing polluted its catchment.
Never have I tasted water as delicious, cool and
refreshing.

Obtaining water while travelling

Most small towns have public toilets, a public
park or recreation reserve, and a service station,
all with tap water that you can use to fill up your
containers (your built-in water tank and 20-litre
drums, as well as any other smaller containers).
If the supply is unsafe, there's usually a sign
labelling it as such. It's advisable to fill caravan
and motorhome water tanks with town water
only, because its added chlorine is beneficial in
keeping the stored water free of bacteria. On the
other hand, though, some people believe that it's
unwise to trust water from anywhere except their
hometown or other major centres. The bottom line
is, if you don't feel confident about the quality of
a water supply, don't take any chances.

Locking your water tank
While travelling, it's a good idea to lock the cap
on your caravan or motorhome's water tank. This
will prevent any form of contamination (accidental or intentional) to your water supply.

WATER TREATMENT

Filtering water

Whatever its source, all water benefits from filtra-
tion. Filtering helps reduce limescale (temporary
hardness) and chlorine; aluminium, copper and

lead; certain pesticides, such as lindane; and various organic impurities like algae, sand and silt. By reducing the above, the taste and odour are improved, and safer, clearer water is the result. Steam-sterilised filter cartridges need to be changed every four weeks or so, depending on the quality of water you're using. However, water filters do not effectively remove giardia, cryptosporidium, bacteria or viruses. Further treatment is required to eradicate these micro-organisms from water.

Purifying suspect water

Experts differ in their recommendations regarding the treatment of suspect water. Most agree, however, that bringing water to the boil and then taking it off and cooling it is *not* sufficient to kill all harmful micro-organisms. The following offers general advice for the treatment of different types of water.

Rainwater from a tank

If you have access to tank water, you may choose to fill water containers with this supply. But remember, not all rainwater is safe to drink. Is there a lid on the tank? Could there be a drowned possum or dead rat polluting the tank? Is the locality aerially sprayed using pesticides? Does the roof catchment look clean? Is there a strange taste, colour or odour to the water? Before using any suspect tank water for drinking or food preparation, it's recommended that you treat it in the following way:

- Using a portable water filter or jug with a sub-micron filter (micron is the size of the particle or solid that will pass through the filter medium), filter the water. This removes algae, silt and sand that bacteria and other harmful micro-organisms feed from and are commonly attached to.

- Boil, at a rolling boil, for at least one minute. Allow the water to cool and then store in a clean container with a lid in your cooler or refrigerator for up to three days. When treated this way, rainwater should be safe for drinking, cleaning teeth, gargling, making ice, washing salad fruits and vegetables, cooking and washing your face and hands.

River, stream, lake, well or bore water

- Using a portable water filter or jug with a sub-micron filter, filter the water to remove algae, silt and sand.
- Boil continuously for a minimum of ten minutes. A campfire kettle is ideal for this purpose. Allow the water to cool.
- Re-filter.
- Store in a clean, sealed container – preferably in a cooler or refrigerator – for up to three days. When treated this way, the water should be safe for drinking, cleaning teeth, gargling, making ice, washing salad fruits and vegetables, cooking and washing your face and hands. If brackish or bore water is used, the taste will remain the same.

Other methods of purifying water

Iodine or chlorine-based purifying tablets can be used if filtering, boiling and re-filtering is not possible. Carefully follow the manufacturer's instructions, especially in terms of the quantity of chemical used and the amount of time you allow the water to stand after treatment.

Desalination is the only option where rainfall is very low. This is a very inexpensive process.

Thermal desalination involves heating water to its boiling point to produce and collect water vapour. Desalination units are available, or you can construct a still.

The ultraviolet treatment of water kills micro-organisms, but has no effect on chemical contaminants. An ultraviolet treatment unit is very expensive to purchase.

SWIMMING

It's important to take care where and how you swim or soak. If you have any doubts about the quality of swimming water or spa water – but feel you *must* swim or soak – don't splash, and take special care not to get water up your nose, in your mouth, or in your eyes or ears. With a raft of very unpleasant and potentially fatal illnesses directly connected to water contamination, this simple policy of keeping your head out of the water makes good sense.

Below are some of the possible health risks associated with swimming:

- Eye infections may result from using a swimming pool that's badly managed, with inadequate amounts of chlorine used. On the other hand, chemical eye irritation may occur if there's too much chlorine used in a pool.
- Potentially harmful waterborne micro-organisms (such as giardia, cryptosporidium, E. coli, blue-green algae and amoebae) may be present in lakes, rivers and badly managed swimming pools. Natural thermal mineral pools can also pose a risk.
- Ears can be invaded by bacteria and fungi,

caused by moisture entering the ear canal. Wearing ear plugs avoids this potential problem.

- Fungal or yeast infections of the skin (especially the feet) may be picked up either by direct contact or by sharing clothing or towels. Poorly maintained spas, containing high levels of bacteria, can cause skin infections.

When to avoid swimming

After heavy rain, avoid swimming at bay beaches and in rivers for at least five days. Whenever there's heavy rain, rivers and other waterways are flushed with contaminated water from gutters and stormwater drains. This water is likely to contain pollutants from street litter, dog faeces, sewage pipe overflow, run-off and seepage from septic tanks, and run-off from upstream grazing pastures polluted with faeces and urine from cattle, sheep or other livestock.

Travellers who enjoy kayaking after heavy rain need to keep the risk of leptospirosis in mind. Leptospirosis is a serious bacterial infection of the liver and other organs. The bacteria (found in the urine of an infected animal) enter a person through a minor abrasion or by being swallowed. Kayakers are the 'canaries' in this case, contracting this serious and sometimes fatal disease. Leptospirosis is just one of many possible water-borne infections.

The high E. coli count in Melbourne's Yarra River in 2005 – caused by leaking sewer pipes and sewer overflows from illegal stormwater connections – shows why swimming and other water sports are risky activities after heavy rain.

Non-polluted sea water contains a large number of

bacteria, although most are harmless to people. On the other hand, an open wound (for example, an ulcer or sore) can become chronically infected by contact with the bacteria present in sea water.

From Robin's travels

While camping in Victoria's western district in the drought of 1967, we decided to swim in a dam instead of using precious drinking water to shower. The dam water wasn't all that enticing, but in we plunged, enjoying ourselves in spite of the mud. Eventually we emerged, only to discover that we were not only dripping with muddy water but with the largest yellow- and black-striped leeches I've ever seen. Barely quelling my feelings of hysteria, I clambered through the mud to the dam bank and the remains of our lunch. Liberal coatings of salt shrivelled the leeches' bloated bodies and they fell to the ground. After that, I really needed a shower, drought or no drought.

Chapter 5
Energy Sources

When camping, you have the opportunity to choose between different forms of energy to supply your energy needs. By using a blend of energy sources for light, heat, hot water, and the storage and cooking of food, you'll have the flexibility to choose the most environmentally sound option for each and every situation.

FORMS OF ENERGY

Solar energy

Sun-powered electricity, harnessed by solar panels mounted permanently on the roof of your caravan, can run many of your energy-efficient appliances. On sunny days, three 60 to 80 watt solar panels will supply enough energy to run an energy-efficient refrigerator, a small TV and a few lights. The panels will also keep your car batteries charged.

To maximise the amount of 'free' energy collected, face the panels directly into the sun, or directly north. A combination of fixed and portable panels provides the ideal system. If you want to park your caravan in the shade, portable solar panels can be carried and positioned in a sunny place for charging.

The maintenance of a well-designed solar energy system is not labour intensive at all. Clean, silent and dependable, it only requires its batteries to be topped up with distilled water every two to three months. The solar panels themselves need to be washed every now and then using clean water, or a little biodegradable cleaner if bird droppings prove difficult to remove.

Generators

It's become increasingly common for campers to use generators to provide power for their appliances. If you need a generator, make sure you purchase the quietest generator possible. Then, in terms of noise and fumes, position your generator so that its effect on other campers is minimal.

Keep in mind that people travel to experience the natural environment, not to hear the hum of a generator. Therefore, never run a generator early in the morning or after 10 p.m. In relation to fire safety, position your generator in the centre of an area of ground cleared of all flammable material. A 1.5-metre clearance is recommended. As well as thinking about other campers, spare a thought for wildlife and domestic livestock.

Gas

Bottled LP gas, which can be used for cooking, heating or lighting, provides an environmentally sound alternative to a campfire or mains electricity. Care must be taken, however, that potentially lethal carbon monoxide does not collect within the small, enclosed space of a caravan, motorhome, annex or tent. A gas detector should be installed to warn travellers of the presence of deadly carbon monoxide, as it is odourless and tasteless.

Gas vapours need a place to escape. In a caravan or motorhome, there must be a properly installed flue as well as good through-ventilation. Two permanent air vents, positioned at either end of the living space, will provide through-ventilation. Additional fresh air can be obtained by opening two windows or one window and a door.

As well as gas vapours, steam (which creates ideal conditions for the growth of mould and mildew) and cooking vapours pollute the air when cooking meals within a small space such as a caravan or annex. These commonly cause respiratory problems, headaches and nausea. When using a stove run on gas or methylated spirits outdoors, check it doesn't outgas harmful exhaust fumes into your space or any other person's area.

Efficient, well-maintained gas appliances are essential. Before leaving on your travels, check gas cylinders for rust, damage or leakage; check hoses for blockages or perishing; and check connections for cleanliness and correct fitting. When not in use, and while travelling, gas should always be turned off at the cylinder.

Never use a gas appliance designed for outdoor use indoors, be it a barbecue, stove or heater. The possibility of carbon monoxide poisoning, as well as smoke inhalation and the chance of starting a fire, means that using these appliances indoors is dangerous. It's also important to remember that you must never use an indoor gas oven or cooker as a heater.

If you have a portable gas or fuel stove, be aware that no naked flame is allowed outdoors on days of total fire ban.

Campfires

Bush camping is an enjoyable activity that many people organise with a group of friends. Enormous campfires are, however, inappropriate. To boil a billy for tea or coffee takes only a few minutes. Build a quick fire using kindling and small pieces of dry wood, then sit back and toast some bread on the coals while you wait for the billy to boil. A small fire is all that's needed to cook a meal and sit around prior to bedtime.

Obeying fire restrictions
If in doubt about the safety of lighting a fire, don't light one. Use a fuel stove instead, but never leave a stove unattended when using it outdoors.

Always check local fire restrictions and regulations with the Country Fire Authority before lighting a campfire. Fire bans operate from November to April in many areas, but vary from one place to another. On days of total fire ban, announcements are made on TV, local radio and on roadside and community signs. No fires (including solid fuel barbecues or ovens, gas barbecues, gas lanterns or fuel stoves in tents) may be lit on days of total fire ban. Before lighting a campfire, it's your responsibility to check the fire restrictions applying to the area in which you're travelling and camping. Heavy penalties exist for anyone found breaking the rules. Ignorance is no excuse.

Constructing a campfire

Campfires are, to many people, an essential aspect of camping. When planning your campsite, position your fire down-wind of your camp so that smoke and sparks don't envelop your outdoor living space. Thoughtful travellers will also locate their fire with other campers in mind. Smoke, although attractive when viewed through the dappled sun and shade of the Australian bush or when rich with the aroma of sizzling sausages, steak and chops, is toxic when breathed in.

Always use fireplaces when they are provided and be content with a small fire. The firewood of inland Australia is well known as dense wood that burns with incredible heat, so a small fire is all you need. Make sure that an adult watches over the fire at all times. Ensure that all combustible material is removed to a distance of at least three metres around and above the fire. Do not allow wood to protrude from the fireplace by more than 30 centimetres, and do not light a fire on a very windy day.

In areas where not many people camp and where there are no fireplaces, it's usually okay to dig a small trench about 30 centimetres deep for your fire. The fire should be no more than 30 centimetres square in size. Clear the site of flammable material for a distance of three metres around the campfire. A fire should not be lit within 7.5 metres of any log or stump.

Have sufficient water on hand in case your fire gets out of control. Some camping areas provide simple, movable campfire drums that protect the grass beneath and don't use as much firewood as a traditional campfire.

Collecting firewood

When gathering firewood, choose only fallen deadwood and twigs lying on the ground, taking care that native plants and animals won't be disturbed. It is preferable to buy or bring your own firewood, rather than alter or destroy the habitat in which many creatures live. To protect firewood from rain, stack it beneath your vehicle or caravan.

Never fell or damage a standing tree, dead or alive. Dead, standing trees are the home and nesting sites of many native animals and birds, including possums, gliders, parrots and owls. Dead trees also provide important perching places for flocks of birds. Hollow logs may be tempting to pick up for your campfire, but before you bend down to pick up a hollow log, pause a while. This log may be home to a number of bush creatures: lizards, snakes, frogs, insects or even small mammals.

From Robin's travels

The river red gum was so massive, and had such presence, that we sat in awed silence beneath its canopy, resting our backs against its trunk. A life force flowed through it: I felt its strength and immense calm. I was more than content. We sipped our coffee, soaking up the stillness, until a Commodore arrived. Out stepped a young couple. We exchanged pleasantries, but were completely taken aback when the man commented, 'Boy, would I like to get a chainsaw into this giant. Think of the firewood. Imagine a slice, right through. What a great dinner table!' Visions of screaming chainsaws, swirling sawdust and blood-red timber filled my mind. A vision of destruction and the collapse of vibrant life. In the retreating dust of the Commodore I felt the sense of wholeness return.

On the other hand, collecting firewood from the ground can assist in reducing an excessive build-up of debris that could lead to a disastrously intense fire which wipes out an entire habitat. This is especially relevant in areas where there are no regular, low-intensity control burns. Certainly, we need to leave sufficient dead timber as nesting sites, food sources and protection for native animals and plants, but perhaps we should relax a little in terms of banning the gathering of firewood altogether in all regions.

Leaving a small stack of firewood (heavy as well as light kindling) beside the fireplace to welcome the next campers to the area is a gesture worth considering.

Leaving your campfire
Before leaving your campsite, make sure your campfire is totally extinguished using water, not soil: if it's cool to touch, it's safe to leave. Flooding the fire trench with water is the only sure way to put out a fire. If you cover a fire with dirt or sand only, the coals will stay hot for a long period of time, creating the potential to burn other people and animals. When your fire is completely out, fill the trench with dirt or sand and replace any grass, leaf litter or stones on top. It's a challenge to leave a campsite looking as untouched as when you found it.

Modern attitudes
Attitudes have changed regarding campfires. Browsing through the bushcraft sections of old Girl Guide and Scout books is a real eye-opener in regard to changing bushwalking practices and philosophies. No longer is it politically correct to remove green branches from trees in order to

construct billy-holding structures over a campfire. Neither is it acceptable to remove dead branches from trees as firewood – nor to pick up from the ground every piece of fallen timber. Modern camp etiquette encourages campers to bring along their own firewood, or use a fuel stove for cooking. Many service stations sell small, convenient bundles of firewood. Alternatively, you can buy artificial fire logs made from recycled, compressed paper waste.

Many popular bushwalking destinations now insist on fuel stoves rather than campfires: for example, the Cradle Mountain–Lake St Clair National Park. This area contains extensive beds of peat that can catch fire and smoulder underground for long periods of time. High use of the area is another reason for the campfire ban.

Modern fuel stoves are light, clean, efficient and quick to use. You don't get streaming eyes or smoke-impregnated clothes. On the other hand, there's always the pleasure of gazing into the coals of a fire and letting your creative thoughts flow. A cheerfully blazing fire seems to encircle and focus the warmth of friendship.

USES OF ENERGY

Most travellers use a combination of energy systems, depending on the availability of mains power, firewood, LP gas and sunshine. Using energy-efficient appliances is always a good policy: select a top-rating fridge, kettle, freezer, microwave and so on. This way you'll make the most of the energy you consume.

Storing perishable food

The storage of perishable food is no longer a challenge; salted, smoked, dehydrated and canned food need not make up your entire menu. These days you can travel into the outback equipped with a small refrigerator and freezer that will keep fresh and frozen food safely for weeks on end.

Chest-style refrigerator or freezer units are the most energy- and space-efficient. Given that cold air sinks, having the door on top reduces the amount of refrigerated air that escapes when the door is opened. However, they're more difficult to pack and unpack, and inevitably the food you require is at the very bottom. Vacuum-sealed, frozen meat lasts best. Foods such as cheese, and fresh vegetables and fruit, have their life greatly extended when refrigerated.

Three-way refrigerator or freezer units run on 12 volts (via a battery charged with energy sourced from solar panels, your vehicle engine's electrical system, or mains power), 240 volts (provided by a generator or mains power) or gas.

Position the refrigerator away from direct sunlight. Make sure that refrigerators and freezers have plenty of ventilation around their perimeter, as well as above and below.

There are now coolers on the market that, although quite expensive, are incredibly efficient and use no power at all. Able to last for up to one week per purchase of ice, these coolers are ideal for keeping most types of fresh food.

Cooking over a campfire

Using the glowing coals of a campfire, foil-wrapped meals are an easy way of cooking.

Care must be taken, though, not to leave pieces of silver foil litter in the campfire itself, or on the ground. This litter is unsightly and will not rot away.

Cooking utensils

Campers use a wide variety of saucepans, kettles, billies and frying pans. These can be made from:

- stainless steel (expensive, easy to clean, but sometimes not thick enough for cooking over a campfire)
- aluminium (light, doesn't shatter, but can melt if overheated on a campfire)
- cast iron (heavy, cheap, has good heat distribution, may rust and will crack if subjected to something cold when it's hot)
- enamel (doesn't rust, easy to clean and suitable for campfire cooking).

A set of billies, which fit inside one another, is a space saver. Black marks on your clothes and hands will be greatly reduced if all your campfire cooking utensils are stored in individual cloth bags.

Whenever cooking over an open fire, the perfect utensil has to be a camp oven. It seems to bring out the very best of flavours. Maybe it's the fresh air, or perhaps there's something about cooking on an open fire with a cast iron pot. Whatever it is, food cooked in a camp oven is delicious.

Cooking with a camp oven

The following tips will ensure you achieve the best results from your camp oven:

- When using a camp oven to bake or roast, place a wire rack in the bottom so that the heat circulates inside the oven. By doing this your food will be cooked evenly all the way through and will not burn on the bottom.

- If you haven't got a rack, don't put hot coals under the camp oven. Rather, place hot coals around the sides of the camp oven and on top of the lid. In order to cook roast meat all the way through, you may need to turn it over.
- Most people sit their camp oven directly on a bed of coals. To allow air to flow freely beneath the camp oven (to maximise even heat distribution), it's a wise policy to use a camp oven with legs, or to place some small stones beneath the oven to allow air and heat to flow properly.
- Another idea is to line the hole with aluminium foil, then shovel in the hot coals prior to positioning the camp oven. Aluminium foil is very useful in preventing heat from escaping into the ground.
- By using a wooden or plastic spoon instead of metal, you will protect the vegetable oil coating within the oven.
- If you need to brown a cake, heap hot coals in the centre of the lid, with just a few around the outside of the lid.
- Realising that different types of wood burn with differing amounts of heat, you will need to adjust your camp oven accordingly. For instance, when using red-gum coals, you'll need just one spadeful of hot coals beneath the camp oven, with two on top. When using wood with less heat, you may need three spadefuls on top.
- When cooking a dish taking four to five hours or more, you'll need to replace the first coals with some fresh hot coals.
- Keeping the lid on will prevent ash polluting your food.

- The lid in itself can be used as a very effective frying pan due to its ability to absorb and retain heat. Simply place it upside down on a bed of hot coals. You can use some small stones beneath the lid, to keep the lid in place.
- To prevent a large piece of meat from burning on the top, simply place a sheet of aluminium foil on top of the roast. Then, to brown the meat, remove the aluminium foil for the final ten minutes or so of cooking.
- If the coals are not very hot or are losing some of their heat, place aluminium foil (shiny side down) over the top of the camp oven and coals. This will reflect heat back into the camp oven and help brown the food.
- Wind can be a problem when using a camp oven, with food cooking on one side only. To get an even heat, place your camp oven in a shallow hole, then place stones or bricks on the windward side, to break the wind.
  Alternatively, wrap aluminium foil (shiny side in) around the camp oven and over the top. This will concentrate the heat into and around the camp oven – even if it's very windy.

Caring for a camp oven

After washing your camp oven and drying it thoroughly, rub the inside using a clean piece of cloth or kitchen paper moistened with vegetable oil.

When putting away your camp oven, leave the lid ajar until completely cold. This will prevent condensation forming that leads to rust.

A camp oven is made from cast iron and is not as tough as it looks. In fact, if dropped onto a hard surface, it will crack.

For the same reason, it's not a good idea to put a very hot camp oven into cold water. The sudden temperature change may cause it to crack.

Store your camp oven in a cloth bag. This will stop the blackened outside from marking your clothes and hands, as well as other equipment.

Using a campfire for warmth

If you're cold and wet, one of the most efficient ways to warm yourself up and dry your clothes is to light two small fires and then sit between them. Even if you're not wet, a campfire provides warmth, especially in the chill of an outback evening or early morning.

From Robin's travels

While travelling in central Australia in an old VW van, we experienced the below-freezing temperatures of the inland. Every evening we warmed ourselves between two small campfires. Although reluctant to leave our twin fires, fatigue eventually drove us indoors and to bed. Lack of insulation, combined with the below-freezing temperatures of the inland, made us long for daylight. On rising in the morning, a glass of water positioned on a shelf close to our heads was always frozen solid.

Lighting

For lighting your campsite, a variety of lamps are available. These may be electric, gas or kerosene pressure lamps. An electric lamp powered with a 12-volt recharging facility is a worthwhile investment for long trips away.

Heating water for showering

Depending on your location, the following are the main methods of heating water while bush camping.

- A campfire will heat sufficient water for a quick shower.
- A solar water bag, usually made of black PVC, uses direct sunshine to heat water for a simple bush shower. Before using this type of bag, rinse it with one tablespoon of bicarb soda dissolved in warm water. This helps cleanse the inside of the bag of chemical residues, especially if you do it a number of times. Using solar energy alone, water can be heated to 50 ºC in just three hours, so remember to test the temperature before showering. For detailed information about solar power while travelling, see Collyn Rivers' excellent book, *Solar That Really Works!*
- Gas can be used to heat water, either by means of a storage unit or as instantaneous hot water. A portable, instantaneous LPG hot-water system (enough to run one shower head) is readily available. If you're camping beside a lake or river where water is abundant, such a system works very well, although it must not be used in an enclosed area due to the danger of carbon monoxide poisoning. Relatively simple and inexpensive to operate, it makes long-term bush camping an attractive proposition.
- In the outback, hot bore water is often the only water available. The Great Artesian Basin is accessed through a pipe drilled to a depth of about 1000 metres, tapping into this vast, pressurised underground reservoir. Frequently the water is hot. Often it's 'smelly' – but only because of mineral content.

- Hot-water units are available that work on a heat-exchange principal, and are connected under the bonnet to the vehicle's radiator. Your vehicle must be stationary, with the engine running. This system relies on the excess heat generated by the engine. The main disadvantages are that the engine generates exhaust fumes and noise.

From Robin's travels

While travelling through central Queensland, the presence of abundant hot artesian water captured my attention. How amazing to need no hot-water system at all, but to simply turn on a tap and have hot water flow out, perfect in temperature for domestic use. On meeting and talking with the locals, we noticed that all the white-haired elderly people of the area had pale orange hair. Dyed through a lifetime of washing in iron oxide-rich (as well as other minerals) bore water, the colour is quite attractive.

At Moree – the artesian spa capital of Australia, located in northern New South Wales – travellers can soak in hot artesian mineral water that has been used since 1895 for its healing and therapeutic value. Rich in sodium carbonate, sodium chloride and other minerals, this thermal water's temperature is 41 °C.

A word of warning: Some travellers scald themselves badly by plunging into geothermal spring water that's too hot. Always test the temperature before entering the water.

Chapter 6
Green Housekeeping

When you think of holidays and travel, you don't think of housework! However, in order to control bacteria, mould and mildew and plain old dirt, certain routines need to be maintained. Good health is essential if you want to have fun. So, while you clean up, think of mountains to be climbed, trout to be caught, surf to be played in, or wildflowers to be photographed. Then feel happy that your travelling home is tiny as compared to your permanent home.

GREEN SHOPPING LIST

While travelling aim to use products that are both people-friendly and environmentally safe. We need to move forward, beyond the use of cleaning products that contain chlorine, ammonia and phosphate. There are green alternatives, solutions to everyday tasks that are:

- effective
- quick and easy to apply
- inexpensive
- and above all else, healthy for people and healthy for the environment.

Generally speaking, if you can safely eat it, the product is also environmentally kind as a cleaner.

Bicarb soda (sodium bicarbonate, baking soda)
Bicarb soda occurs abundantly in nature but depends upon sophisticated processing in order to meet high standards of quality and purity. This non-poisonous powder is inexpensive and has a multitude of uses. It will remove stains, absorb unpleasant smells, soften water, relieve itching and can also act as a toothpaste and a deodorant.

Borax
Borax is a naturally occurring mineral salt. This fine white powder acts as a stain remover, grease solvent, fabric and water softener, and soap booster. It also possesses bleaching and disinfectant qualities. Borax (when used as a powder) is effective in controlling insect pests such as cockroaches and silverfish.

It must be recognised that borax is poisonous when swallowed, so care needs to be taken in its use and storage. It can also enter the body through broken skin.

Cloves
Cloves are the dried flower buds of tropical myrtles, used as spice. They are non-poisonous, sweet-smelling, readily available and inexpensive. Clothes moths find them totally unacceptable.

Eucalyptus oil
This natural oil is distilled from gum trees (eucalypts). It has many medicinal properties ranging from relieving the symptoms of colds and influenza, to easing aches and pains. Eucalyptus oil is a penetrating oil that evaporates rapidly and is

useful as an antiseptic, disinfectant and deodorant. It can be purchased as a spray or liquid.

As a cleaning agent, its uses range from freshening a load of washing; to removing grease, gum and stubborn stains from clothes; to lifting tar and adhesive material from paintwork and dog's paws! It is also an important part of most wool washes and is useful as an insect repellent.

Although natural in origin, eucalyptus oil contains ingredients that are highly toxic when swallowed. Keep out of reach of children.

Flyswats
One for the caravan, motorhome or campervan and you can get rid of that pest strip and all your aerosol sprays.

Pure soap
Dependable, durable and made of ingredients that are environment-friendly, pure soap is unlikely to cause skin irritations or allergies. It makes grease, food residues, bacteria and all other forms of dirt disperse in water.

Salt
Salt is naturally occurring and non-toxic. To use as an antiseptic to clean cuts and grazes, simply dissolve half a teaspoon of cooking salt in one cup of boiled water. This solution is recommended by dentists as a mouthwash to treat gum disease and also after tooth extraction. As an eye bath, eye specialists recommend a weak saltwater solution.

Salt is also a valuable product to clean and disinfect food preparation areas and utensils. Its uses are widespread.

Scourer (nylon) and steel wool (plain, fine)
A simple, inexpensive and highly effective means

of removing stubborn dirt without using powerful and potentially harmful cleaning products.
River sand or beach sand are handy abrasives to
use when camping and can be used instead of
detergent.

Vinegar (white)
White vinegar is an inexpensive, clear, colourless
liquid made by fermenting cereals or cane sugar.
This mild, natural acid (acetic acid, with an
acidity of about 4 to 5 percent) is able to neutral-
ise grease and soap residues. It is also an
anti-mould agent, a mild disinfectant, a bleach
and a deodorant. Vinegar is the very best general
purpose cleaner.

Washing soda (sodium carbonate, 'Lectric' soda)
This crystalline powder (or crystals) is useful as
a water softener, stain remover and degreaser –
especially for blocked drains. It can also be used
as a poultice to reduce swelling and in a basin of
warm water to relieve aches and pains.

Laundry and kitchen detergents
Purchase brands that are:
- *Biodegradable* (capable of quick decomposition
  into ecologically harmless products).
- L*ow phosphate* or *phosphate-free*. The problem of
  declining water quality in our inland rivers,
  lakes and waterways is a serious ecological
  issue. One of the causes of this problem has
  been identified as an excessively large quantity
  of plant nutrients entering the waterways, especially phosphates. It has been estimated that 30
  to 50 percent of all phosphates that pass
  through sewage farms come from the use of
  domestic detergents and other cleaning products. You can play your part in helping to

reduce this pollution – even when travelling –
by making small changes in your routines.

USING GREEN CLEANERS

Water (hot, warm and cold)		Vinegar (white)	Lemon juice	Eucalyptus oil
Soap (pure)	Bicarb soda	Salt	Borax	Washing soda ('Lectric' soda)

Working with a soft cloth, nylon scourer or plain,
fine steel wool, you can combine any of the liquids in the top row, with any of the solids below.

These green cleaners increase in strength, from
left to right. Warm, soapy water, for instance, is
a very effective way of cleaning most surfaces.
Stubborn stains sometimes require borax, 'Lectric'
soda or eucalyptus oil.

GREEN KITCHEN

Your caravan can be kept attractive and sparkling
clean using green, environmentally friendly products – with harmful bacteria kept at safe levels –
such as plenty of soap and hot water, vinegar, salt
and eucalyptus oil; as well as nature's sterilisers,
frost and sunshine. Remember: heat, cold and
dryness all help in controlling harmful bacteria.

Washing your hands in warm, soapy water is an
extraordinarily important habit to adopt after
going to the toilet, and before eating or preparing
food. Research suggests that ordinary soap elimi-
nates harmful bacteria as effectively as antibacterial soap. The aim is not to sterilise, rather to
reduce the number of harmful micro-organisms to

a level where they will not cause illness. Ensure, therefore, that soap and a hand towel are always available for easy use. Dry your hands thoroughly before touching food.

Dishes should be washed in very hot soapy water. Soap helps kill bacteria and viruses. After washing dishes, make sure they are completely dry before putting them away. Bacteria don't like being dry.

Barbecue cooking plate
Sprinkle salt (to absorb grease and dirt) on the barbecue plate while it's still very hot, then leave to cool before brushing clean. Use scrunched-up newspaper to wipe away the last bits of grime. The paper can then be burnt in your campfire. Protect from rust by applying a thin film of vegetable oil. By using cleaning products that are so safe you can and do eat them, you can be sure that your next barbecued steak isn't marinated in harsh cleaning product.

Chopping boards
It's very important that chopping boards are thoroughly cleaned after use, especially after cutting up raw meat. Scrub plastic with hot soapy water. Rinse well, wipe over with vinegar, then allow to dry.

To clean wooden boards, rinse with cold water then rub in plenty of salt. Hot water opens up the grain and allows germs and odours to penetrate the wood. When dry, season the wood with a little vegetable oil. Keep one chopping board for raw meat only, another for vegetables to be cooked, and yet another for salad fruits and vegetables.

White polythene chopping boards can be rubbed clean using a paste made up of equal parts salt and bicarb soda, with water. For bad stains, add lemon juice to the paste rather than water.

Half a lemon rubbed into the surface will clean the board of onion, garlic and fish odours.

A final hint: develop the daily routine of hanging your dishcloth, scourer and mop in the sunshine to dry and let nature do the sterilising.

Cleaning (general)

Since the kitchen is where you prepare food, a clean kitchen is essential. It need not, however, be at the expense of the environment.

The following daily tasks will keep dust mites at acceptable levels, as well as mould and bacteria, ants, cockroaches, flies and silverfish. Shared between members of your travelling team, this routine is easily managed.

- Use a dry towel to wipe over any wet areas. This will reduce the build-up of soap scum, bacteria and mould, especially if followed by a quick vinegar wipe-over.
- Sweep your eating area floor.
- Do an immediate wipe-up of food spills (and other dirty marks) on benches, tables and floors. This way spills never spread or become difficult to remove.
- Check that food is correctly stored in your freezer, refrigerator, cooler and in sealed containers in your pantry. Stale food, food crumbs and residues will attract and feed bacteria, cockroaches, flies and ants.

Hot soapy water is sufficient, in most cases, to clean kitchen surfaces. Bicarb soda is a gentle cleaning agent. It can be used as a powder, paste or solution. Stains on all types of surfaces can be removed using bicarb soda as a paste and applying it with a damp cloth. Rinse well, using hot water. Stainless steel surfaces, enamel, stained tea and coffee cups, refrigerators, china teapots, laminex surfaces and plastics all clean brightly with an application of bicarb soda.

If the stain is more stubborn, cover with a layer of bicarb soda, then sprinkle with vinegar. While bubbling, rub, using a soft brush, scourer or cloth. If you leave the bicarb soda and vinegar to sit for ten minutes, the stain will soften. Rub clean, then sponge off the bicarb soda. Finish the job with a vinegar wipe-over.

Vinegar can be used anywhere in your travel set up: on benches and tabletops, windows, floors, stainless steel sink and plastic basins, toilet seat and in your rinse water for clothes and dishes.

Salt is ideal for cleaning sinks and chopping boards used for food preparation, due to its excellent cleaning and disinfecting qualities. Glass, marble, metals and laminex also clean well using salt. Combine bicarb soda and salt, to make a tough cleaning powder.

If you've become accustomed to using a commercial spray-and-wipe, you may like to pour vinegar into a spray bottle for use in a similar way. Alternatively, dissolve bicarb soda in hot water, add vinegar and use as a spray.

Cockroaches

If cockroaches are a problem in the area, the following routine is especially important:

- During the hours of darkness, leave your food preparation area as clean as possible. Your aim is to provide as little food and water as you can for these unpleasant insects to feast on.

- In the morning, give your benches, sink and stovetop a quick wipe-over with white vinegar, to ensure clean food preparation surfaces.

Cooking smells

When cooking foods such as fish, a bowl of vinegar placed beside the pan will help reduce the smell of hot fat and oil. If the smell is still strong, add one tablespoon of vinegar to the washing-up water.

It can give you satisfaction to catch a fish for dinner, but you don't want to smell fish every time your hands are anywhere near your nostrils! To remove unpleasant or strong odours from your hands as a result of preparing foods like onions and fish:

- Rub hands with salt and water, or bicarb soda and water.
- Or, work the juice of half a lemon into your skin.

Cooking utensils and plates (greasy)

Cooking utensils and plates that are particularly greasy can be wiped with newspaper or a paper towel before washing up. The paper can then be burnt in your campfire. By dealing with oil or grease in this way, you avoid the use of large amounts of detergent. Scourers or river sand with hot water may be all that you need to clean dirty dishes. Alternatively, use hot soapy water or a little biodegradable detergent.

Dishcloths

Soak a smelly dishcloth (the smell indicates the presence of bacteria) overnight in a solution of vinegar and hot water. Rinse, then hang in the sunshine and let nature do the deodorising and sterilising.

Drain (blocked)

If your caravan drain becomes blocked, try one of the following green suggestions:

- Pour one cup of bicarb soda down the drain, then follow with one cup of white vinegar. After the mixture has fizzed and bubbled, slowly pour down a jug of boiling water.
- Pour one cup of washing soda crystals ('Lectric' soda) into the drain then follow with a jug of boiling water.

Food scraps

When staying in eco-resorts, it's good to know that your food waste isn't going to be dumped into landfill. A fermentation system – using a bench-top recycling bucket – has been designed using beneficial micro-organisms to ferment organic waste into fertiliser. All you do is open the lid, pour in the kitchen scraps, then add a sprinkling of *bokashi* (a grain impregnated with micro-organisms). After three to four weeks, the non-smelly decomposed liquid contents are emptied via a tap into a bucket. This liquid is diluted with water then used as a liquid fertiliser in the garden. Covered with soil or mulch, this nutrient-rich fertiliser benefits any garden. The unit sells as a kit from Eco Organics in Eltham, Melbourne.

Food storage

There are a number of strategies you can use to minimise both spoilage and spillage of foods:

- Place all foods within strong containers.
- Select containers that fit neatly within cupboards or shelves.
- Pack food so that it doesn't move around within the container.
- Pack foods such as honey, cordial, cooking oil,

jam, dressings and sauces in a strong, leak-proof container. If any of these foods spill, they won't travel through the entire cupboard, making a terrible mess. You'll only need to rinse out one container.

- Ideally, transfer liquids from glass containers into strong flexible plastic containers with sturdy screw-top lids.

- To avoid breakages and spills, pack strong plastic bags filled with items such as sugar, rice or pasta between glass bottles and cans.
Alternatively, place wads of cardboard or paper between cans or bottles to prevent friction-type damage leading to messy spills. Don't pack cans of drink against one another, as these can rub together enough to puncture.

Microwave oven

After use, wipe the inside with a hot damp cloth, especially if there's been a spill. For more stubborn stains, place a bowl of hot water (containing a slice of lemon) in the microwave. Set to simmer until the interior is quite steamy. Leave for a few minutes, then remove the bowl. Now wipe away the stains using a hot damp cloth.

To deodorise, add the juice of one lemon to half a cup of water. Place in microwave and turn on high setting for one minute. Wipe out. Repeat if necessary.

Strong odours produced by foods such as pickles, chutney, fish or onions can be removed by wiping over the oven's interior with equal parts vinegar and hot water. Leave the door open overnight as well.

Net (to cover food)

A simple net that opens like an umbrella is useful to keep flies away from freshly barbecued meat and salads.

Oven

While your oven is still warm, wipe with a damp cloth sprinkled with bicarb soda.

If oven is really grimy, apply a bicarb soda paste to all surfaces, then leave for 30 minutes. Now wipe clean. Use a wooden scraper on any difficult patches. Add vinegar or lemon juice to the bicarb soda for extra cleaning power.

Soak oven racks in hot soapy water and bicarb soda. Polish clean with steel wool, rinse with very hot water and dry with a soft cloth.

To prevent a grimy oven, make up a mixture of one tablespoon of bicarb soda dissolved in one cup of water. After cleaning, 'paint' your oven with this mixture. When dry, the bicarb soda and water will leave a hard, sterile skin over the clean surface. Any fat, oil or burnt food will now stick to the rough surface made by the bicarb soda. The bicarb soda and grime will lift off easily using plain hot water. Every month or so, recoat all the surfaces again. This way you can be certain you won't serve up a film of harsh oven cleaner along with your next baked meal!

When you clean your oven using bicarb soda and vinegar, any lingering smells will be neutralised, especially if you leave on the recommended light coating of bicarb soda and water.

Oven (food spill)

Sprinkle with salt then brush off the burnt food, which will have combined with the salt.

Plastic drink bottle
Soak overnight in a solution of bicarb soda and warm water to remove both the taste and smell of the plastic.

Plastic ware
Plastic ware is unbreakable and therefore ideal for camping. Use bicarb soda dampened with water to clean stains off plastic cups, plates and bowls. This will not damage the surface.

To remove unpleasant odours from plastic containers, soak in bicarb soda and warm water. Rinse well. Alternatively, use lemon juice and water, or vinegar and water. Wipe dry, then store with the lid slightly ajar.

Refrigerator (packing, unpleasant odours and cleaning)
The same rules that apply to food storage on pages 89–90 can be used for the placement of food within your camp refrigerator. In addition:

- When placing foods that might roll around (such as eggs, apples, tomatoes and other fruits and vegetables) in your refrigerator, pack into special containers designed to fit on particular shelves.
- Always check that the refrigerator door is firmly shut and locked, prior to departure.

To mask unpleasant odours in your refrigerator:

- Place half a lemon on a shelf to help absorb food odours.
- Or, leave an open packet of bicarb soda in the refrigerator. This will be effective for about three months.

To keep your refrigerator sweet-smelling and free of mildew, wash the surfaces with a solution of bicarb soda and hot water, or sponge with vinegar on a damp cloth.

Wipe all surfaces, inside and out, with a hot vinegar and water solution. For the sealing strips around the door in particular, add a few drops of eucalyptus oil to one cup of very hot water, then wipe over the sealing strips. By this method, you can control mould and mildew.

Rubbish bin
Make sure that rubbish is always stored in a sealed bin. Use plenty of hot water, with a little detergent and eucalyptus oil added, to clean and disinfect your rubbish bin.

Steel wool
To prevent steel wool from rusting, place the used pads in a screw-top jar in which one teaspoon of bicarb soda has been dissolved in warm water. Alternatively, use a soapy mixture.

Tea and coffee stains
Rub a bicarb soda and water paste into stains on china, ceramic or plastic cups and mugs. Rinse off and dry with a soft cloth. Bicarb soda will not scratch the surface.

Thermos
A stainless steel vacuum flask is a worthwhile investment. Filled with boiling water after breakfast and lunch, you'll have plenty of hot water for morning and afternoon tea breaks.

To freshen up your thermos, place one teaspoon of bicarb soda and warm water in the flask and allow it to soak for a few minutes. Rinse well.

Store a piece of charcoal in your thermos to keep it smelling fresh. Remove before use.

Vegetables
To help remove pesticides, bacteria and traces of
mould or mildew from vegetables (both salad
vegetables and vegetables to be cooked), combine
a little vinegar with water and rinse. Brush clean.

Vinyl (scuff marks on)
Use eucalyptus oil to rub away the marks.

Vinyl floors (also linoleum)
Mop using a solution of one cup of vinegar to half
a bucket of very hot water, or rub away stains and
other dirty marks using a cloth moistened with
vinegar. The vinegar cleans, deodorises, is an anti-
mould agent, prevents spotting as the floor dries
and leaves a shiny surface.

Water tanks (built into a caravan or campervan)
If the water becomes smelly, put four tablespoons
of bicarb soda into the water tank, take your
campervan or caravan for a short drive, then
drain the tank. The bicarb soda will help remove
any slime or unpleasant smell.

GREEN LAUNDRY

The contamination of our environment has
become an important issue and one that concerns
every thinking person. You may think that the
quantity discharged from your caravan is small,
but the cumulative effect of many thousands of
travellers is drastic. By making small changes in
your domestic routines, you can significantly
reduce the pollution in our rivers, lakes and
oceans.

As responsible travellers, it's up to us to ensure
that our waste water is as biodegradable as possible and contains no phosphates. We can feel

confident of this if we use the following hints and products.

Camp-style washing machine

A camp-style washing machine is easily made. Half fill a plastic tub with water, then add a little biodegradable detergent and your soiled clothes. Clip on a watertight lid. Wedge the bin in a corner of your caravan or vehicle. The bumps, corrugations and other movements will wash your clothes as you travel. When you arrive at your campsite, rinse the clothes and hang them to dry.

Felt hats (Akubra)

To remove spots, rub lightly with fine sandpaper, or sponge or spray with eucalyptus oil to remove oil, dirt or grease stains.

Grease or oil (on porous surfaces)

Camping invariably leads to the occasional grease (or oil) stain. For any porous surface such as carpet, fabric, leather, bricks, concrete or stone, follow this procedure:

- Absorb surface grease with paper towelling. Alternatively, sprinkle with dry sand, then sweep clean.
- Combine equal parts chalk and bicarb soda with enough water to make a sloppy paste.
- Spread paste over the stain. Seal with plastic.
- Leave until completely dry.
- Brush (or vacuum) away the chalk and bicarb soda, which will have absorbed the oil or grease. Repeat if necessary.

Alternatively, pour boiling water on the grease spot, then dust liberally with bicarb soda. Work the bicarb soda into the stain, then wash normally.

Hard water

Hard water, common in outback Australia, can be softened for washing by dissolving washing soda (or borax or bicarb soda) in hot water. With the water softened you can use pure soap instead of detergent or expensive washing powder. By adding vinegar to the rinse water, you soften both the water and the fabric.

Rinsing

To remove all traces of soap or detergent, add half a cup of vinegar to the rinse cycle.

Half a teaspoon of eucalyptus oil will ensure a fresh smell, help repel moths and silverfish, and assist in dust mite control.

Smoke (tainted washing)

If a neighbour's campfire has left your washing with an unpleasant smoky odour, put the articles back in a bucket or washing machine. Now add hot water, with a generous amount of white vinegar. Let the articles soak for 15 minutes, then ring out or spin dry. All traces of smoke will have disappeared.

From Robin's travels

Since I'm not fond of ironing, I give our clothes a good shake before hanging and peg, unpeg and fold carefully. This way I eliminate the task of ironing, especially while travelling.

Recycling old bath towels makes a lot of sense. Simply cut them up into hand-towel and face-washer sizes, then sew along the cut edges. You need plenty of hand towels while travelling.

Wood ash was traditionally used by the Greek community to soften water used for washing. A kilo or more of ash was placed in a cloth on top of the family wash. Boiling water was poured slowly through the ash, then the clothes were left to soak overnight. This can be a fun thing to do while camping. But until you get the hang of it, try the method on garments that don't matter.

Stains (on clothes)

There are many, many green ways to tackle stains. If you can't get the stain out using these methods, then you've got to ask yourself, does it really matter?

With a fresh stain, the rule is: *don't let it dry out completely.* Some dry stains prove impossible to remove, no matter what you do, especially those caused by meat juice, blood, fruit or egg white. Therefore, aim to treat stains while they are fresh. When time is short (or you don't want to make a fuss), simply sponge with cold water, soda water, or a white vinegar and cold-water solution, then leave a damp towel over the stain until you can deal with it properly.

Due to the wide variation of materials used in the manufacturing process, it is always wise to test the solution being used on a small piece of the fabric. To avoid a ring developing, first treat the area around the stain, then work back towards the centre. Remove stains before putting articles through a normal wash cycle. With carpet, avoid over-wetting.

Your green cleaners can be used in the following ways to treat stains:

- Bicarb soda softens fabrics as well as the water and helps remove stains.
- Borax acts as a soap booster, grease solvent and

mild bleach and will dissolve most dirty spots. Wash in a borax and warm soapy water solution.

- Eucalyptus oil or spray can be used to remove biro, chewing gum, grass, grease, gum, glue, ink, lipstick, nicotine, oil, shoe polish, tar or any stubborn unknown stain. Simply place an absorbent cloth under the stain, then dab or spray with eucalyptus oil, working towards the centre of the mark. Follow with a normal wash.
- Methylated spirits can be used to remove biro, felt pen, grass, grease, nicotine or shoe polish.
- Washing soda (sodium carbonate) softens water and removes stains and is excellent at removing grease. For best results, dissolve crystals in hot water, then leave greasy articles to soak before putting through a normal wash cycle.
- Citrus-based cleaners, which use a blend of natural extracts and no phosphates, can be used like eucalyptus oil to successfully remove a wide range of stains.
- If your carpet, rugs or car upholstery has a slight overall greasiness or odour, you can dry-clean it yourself. Simply fill a sieve with bicarb soda (or cornflour), then sprinkle the powder through the sieve and onto your carpet or rugs. Using a soft broom, brush the powder into the pile. Leave it overnight, or for at least half an hour. Finally, vacuum thoroughly and you'll find that the bicarb soda has absorbed all the odour, grease and loose dirt.

GREEN BATHROOM

Air freshener
Commercial air fresheners tackle smells the wrong way. It's wiser to deal with the source of the

problem and then improve ventilation, rather than mask unpleasant smells with artificial fragrances.

Eucalyptus oil (or spray) can also be used as a natural air freshener. Activated charcoal and bicarb soda absorb stale smells.

Camp toilets

Wipe toilet seat daily with a cloth moistened with vinegar to give a clean, shiny, odourless surface. Remove any stains by applying a paste of bicarb soda and vinegar. Rub clean. To keep flies at a distance, spray with eucalyptus oil.

Keep toilet paper in a plastic bag, attached to the inside of the tent. Place a basin of water (ideally on a portable wash-stand), some soap and a hand towel close by for use after going to the toilet.

Caravan park amenity blocks

When you use a caravan park amenity block, you'll notice there's a certain etiquette involved.

- Some amenity blocks leave a radio playing 24 hours per day. Others prefer silence. A radio has the positive effect of drowning out unpleasant 'noises' – a definite advantage when a lot of strangers share one facility.
- Where a caravan park is located close to a town, it's usual for each traveller to be given a key to the amenity block. The aim is to restrict its use to paying guests only.
- Do not presume that other paying travellers share your taste in perfume – remember that some people suffer health problems as a direct result of perfume. Therefore, leave your application of all fragrances (deodorants, powder, perfume, hair spray and aftershave) to the privacy of your tent, caravan or campervan.

- Leave the shower cubicle free of powder, mud and sand. To help you leave the floor clean for the next person, some parks provide a bucket and mop.
- Enjoy your shower but use water wisely, remembering that it's a scarce resource in most areas. Where water restrictions are in place, coin-activated showers encourage travellers to be aware of their water usage. Ten- or 20-cent coins are usually required.

From Robin's travels

A close friend, on hearing that I'd be camping mid-winter, offered me this advice. 'Get yourself some roomy flannelette pyjamas as well as track pants and a top to wear over them. Have your shower in the afternoon, when there's still some warmth in the air. When you go to bed at night, just take off the track pants and top and leap into bed. This way you'll have warmed up your night wear.' I followed her suggestion, especially during frosty weather.

Sponge bath
When bush camping, with no permanent water on hand, an all-over wash using a small basin of water and a face washer is surprisingly effective. (But start with the cleanest places.)

Towels
Hang up your bath and hand towels to dry. Dryness kills both bacteria and mildew.

GREEN LIVING AND SLEEPING AREAS

Develop a 'tidy as you go' habit in your living and sleeping areas. In the small space of a caravan or campervan, tidiness is important for your peace of mind. This way you'll have the pleasure of a well-organised travel outfit and no longer spend large slabs of time trying to locate, for example, your car keys or sunglasses.

Bedspread (as a dust cover)
A dust cover or shiny-surfaced bedspread will keep bedding dust-free. At the end of a day's travel, lift off, give it a shake and place in temporary storage.

Canvas (care of)
In order to keep canvas tents and annexes free of mould and mildew, and in good condition, follow these hints:

- Keep the tent floor as clean as possible (but remember you're on holiday!).
- While you're travelling, try to ensure canvas is dry before folding.
- After your holiday, canvas must be completely dry before folding.
- Store canvas equipment, long-term, off the ground in an airy cupboard.

Mould and mildew (as well as musty odours) are easily controlled by sponging the fabric with plenty of white vinegar. Allow to dry, then rinse with water. Alternatively, rub in plenty of dry bicarb soda, leave for 30 minutes, then brush clean. Mould and mildew will damage canvas if left untreated.

By avoiding the use of fly sprays, detergents, stiff brushes and high-pressure water, you'll protect the tent canvas from damage and discolouring.

However, stains on canvas need not be a cause for alarm. Some will remind you of good times, although if you prefer:

- Red dust can be brushed off using a soft-bristled brush.
- Sap from trees and oil are more difficult, but may be reduced by dusting liberally with bicarb soda.
- Bird droppings can be scraped off when dry, then sponged with a vinegar and water solution.

Cupboards (damp)
When travelling in places with high humidity, mould and mildew can be a problem. Place a few pieces of chalk on the shelves to absorb moisture and prevent the growth of mildew. Spray lightly with eucalyptus oil and improve ventilation. Alternatively, sponge with white vinegar.

Curtain rods
Candle wax or soap smeared on curtain rods will encourage the smooth pulling of caravan curtains.

Curtains and blinds
Dust is an inevitable part of travel; therefore, choose curtains and blinds without dust-collecting frills or ruffles.

Deodorise (small living spaces, cupboards and drawers)
Lingering food odours like garlic or fish and chips, or mould or mildew on carpet – all these and more cause problems in small living areas.

Fresh air is both free and the very best deodorant, so open up those doors and windows! However, if nature needs a little help, use any of the following in a small bowl: vinegar, cloves, activated charcoal, or place an open packet or bowl of dry bicarb soda in the area.

Mobile phone
A regular wipe-over with a little white vinegar on a cloth cleans and disinfects your phone.

Mould and mildew (caravans)

Black, green-black or brownish specks or stains on your walls and ceilings are minute furry growths of fungi – mould and mildew. These thrive in damp, warm, poorly ventilated spaces. They can also live under carpets and in cupboards.

Mould produces microscopic spores and potentially harmful chemical vapours that float in the air. When they are inhaled or swallowed or land on the skin, they cause allergic reactions in some people. Apart from the health problems triggered by mould and mildew, there's also the unsightly aspect to contend with. To kill and clean away mould and mildew, use a strong solution of any of the following:

- White vinegar, salt, eucalyptus oil or tea-tree oil (with the oils, add a few drops to a basin of very hot soapy water). Simply wipe over the surface, leave for ten minutes or overnight, then finish with a vinegar and hot water rinse.
- If the mould is particularly ingrained, combine bicarb soda with vinegar and use a stiff brush to clean the area – and if possible, use sunshine to dry.
- Eucalyptus spray is useful as a preventative measure.
- Remember, mould and mildew thrive in warm, damp, poorly ventilated places, so use an extractor fan and allow good through ventilation.

In hot, humid climates, toothbrushes, shoes, handbags and clothes stored in poorly ventilated cupboards frequently grow mould and mildew. Here are some steps to minimise mould and mildew:

- Always dry shoes, bags, coats or umbrellas before putting them away in a well-ventilated rather than airtight cupboard.
- Store linen and clothing in airy cupboards.
- Wipe over any mould-susceptible surface with vinegar.
- Rinse any mould-susceptible fabric in a vinegar and water solution.
- Give mattresses, underblankets and pillows a regular three hours in sunshine.

Reading glasses and sunglasses

A drop of white vinegar on each lens then a quick rub with a soft cloth ensures a clean, streak-free surface.

Sleeping bags and doonas (airing, cleaning and storage)

Inland Australia is well known for its frosty nights. For this reason, a good quality sleeping bag (preferably with a hood) or doona is an essential piece of outback equipment. Keeping it clean, well-aired and free of mildew is essential. To maintain super down and synthetic doonas and sleeping bags in their best insulating condition, follow these hints:

- Every week or so, take your doona or sleeping bag outside, shake well and then hang in sunshine for three hours. This will give it a real lift in terms of its insulating capacity and a fresh, clean smell.
- Sponge any dirty spots using a vinegar and warm water solution.
- If grubby all over, wash using pure soap designed for the task.
- Rinse in warm water to which you've added white vinegar. The vinegar will remove any

traces of soap and protect the fabric against mould and mildew.

- Squeeze out as much of the water as possible, then dry flat, out of direct sunshine. Turn it over every 30 minutes or so. Shake to fluff up the inside insulating material.
- When your travels are over, store sleeping bags or doonas in a dry, well-ventilated cupboard, in loose cotton bags.

GREEN VEHICLE CARE

Flywire screens on caravan windows
If possible, take your flywire screens outside to brush away loose dust, spiders' webs and insects. Make sure that the wind carries the dust away from you. Using vinegar and very hot water (one cup of vinegar to half a bucket of very hot water), wash screens and clean away any traces of mould and mildew. Give a final wipe-over with pure vinegar, or eucalyptus or tea-tree oil (a few drops in a basin of very hot water). Finally, dry the screens in the sun.

For screens that can't be taken down, use a brush (or vacuum cleaner) to clean away dust, sticky spiders' webs, insects and mould spores. Then wash and wipe over, as above, and allow to dry.

Grease stains
Grease stains on car upholstery can be removed by working dry bicarb soda into the spot, then vacuuming or brushing it away.

Alternatively, eucalyptus oil or spray may be used to remove oil, grease, gum and other stubborn marks from duco or car upholstery.

Odours

Stale cigarette smoke, sweat, vomit, spilled milk, wet dogs, lingering food odours like fish and chips, and mildew on carpet – all these and more cause problems in cars.

When you clean up smelly spills, use plenty of bicarb soda, then rinse with vinegar. An open packet of bicarb soda in your car will absorb unpleasant smells for about three months.

Oil or grease (spilled on concrete or paving)
See page 95.

Plasticisers (that new car smell)

When you first open the door of a brand-new car, its chemical smell will seem either offensive or seductive, depending on the car and your taste. Likewise if your car is relatively new and has been left locked up on a hot day (especially sitting in the sun), the combined heat and the smell can be quite overpowering. Such chemical odours come from the plastics, vinyls, dyes and glues used in the interior trim. It's not healthy to be breathing in these chemicals.

Car producers are trying to reduce the number of chemicals they use on interior finishes. Some locally produced models have moved towards vegetable-based dyes and water-soluble solutions so as to reduce the chemical cocktail given off by the interior of their cars.

If you find the smell of plasticisers in your new car disagreeable, wipe over the vinyl and plastic surfaces with a strong solution of vinegar. Rinse well, then use the wind and the sun to dry the interior of the car.

Alternatively, a solution of bicarb soda and warm

water may be used to wash smelly vinyl seats and lining panels.

Registration sticker (to remove)

To remove the sticker from the windscreen, wet a piece of cling wrap and place it over the sticker. Leave for 30 minutes, then peel it off.

Alternatively, press wet newspaper over the sticker until the sticker comes off easily. Eucalyptus oil will remove any sticky material.

Tar

Moisten a cloth with eucalyptus oil and rub clean. Often it's necessary to let the eucalyptus oil soak into the tar for some time to soften it. A warm sunny day will help with the softening process. You may need to repeat the process several times until the tar comes away.

Tow ball

To stop the problem of grease on clothing from a tow bar not in use, simply cut away about one-third of an old tennis ball, then slip the remaining two-thirds over the tow ball.

Washing outside surfaces

Be aware that in many places water is too scarce to use for washing vehicles. If no water restrictions are in place, park your vehicle on a grassed area that drains into a garden bed, then wash it using an environment-friendly detergent. This will reduce the amount of chemicals and run-off into storm water drains, eventually finding its way into precious waterways and the ocean.

Windows, windscreen and headlights

Add one cup of vinegar to half a bucket of hot water. This washing solution will leave a clean, streak- and spot-free surface. Alternatively, use hot water and an environment-friendly detergent to clean these surfaces. For a very quick job, simply use a soft cloth moistened with vinegar.

Alternatively, sprinkle bicarb soda on a damp cloth and rub clean. Polish dry for a gleaming surface free of squashed insects, bird droppings and traffic grime.

Windscreens (internal)

Over time, the internal windows and windscreen of your car can develop a layer of film due to the release of chemical vapours from vinyls, dyes and glues. To remove this 'plastic' film and help reduce fogging, moisten a ball of newspaper with vinegar and rub clean. Alternatively, moisten paper towelling with eucalyptus oil and rub clean.

From Robin's travels

A lint-free cloth moistened with vinegar is my way of cleaning our car's steering wheel. I've a strong dislike of sticky steering wheels! Whenever we change over to a new or second-hand car, we always wipe over all the internal surfaces with a strong solution of vinegar, to help remove the smell of plastic. Then we air the car thoroughly, allowing the sun and wind to complete the task.

I find that a soft cloth moistened with vinegar does an excellent job on the windscreen and windows. When driving close to the ocean, car windows are often coated with salt spray – as well as dust – in the drier months. We find that a quick wash-over with hot water and vinegar works wonders. Driving glasses and sunglasses receive 'the drop of vinegar treatment'. It works extremely well and is so simple.

Chapter 7
Waste Disposal

As well as the practicalities of travelling green, your attitudes and sensitivities to green issues play an important role. A person who's respectful of other people and the environment will be a traveller the locals feel comfortable with; a traveller who feels a true empathy with the environment in which they stay. Behave as a guest and you can't go wrong. New friendships will flourish in this mind-set, along with enlightenment regarding appreciation of all things natural.

TOILETS

Poor hygiene and toilet habits have polluted many areas throughout Australia. Ignorance is no excuse. Behaviour such as this is disrespectful of the environment, fellow travellers and also the local population. Gastroenteritis (diarrhoea and vomiting) is common where human waste pollutes the ground or the water supply. Giardia, cryptosporidium, E. coli and other harmful microorganisms live in water contaminated with human faeces. Disease inevitably follows poor toileting habits. As responsible travellers we must

take great care to preserve the purity of all waterways, but especially those in delicate ecosystems.

When selecting toilet paper, choose unbleached,
unperfumed, dye-free paper. If using condoms,
sanitary products or disposable nappies, take
soiled products away with you and dispose of
them – sealed in biodegradable plastic bags – in
the next town's tip or rubbish bin. These products
take a long time to decompose.

Public toilets

Always use public toilets if they're provided. It
seems characteristic of mature-aged westerners
that we can't squat easily, and holding a crouched
position for more than a minute or so usually
proves very uncomfortable. Therefore, most of us
prefer a toilet seat, even if it's in a public toilet
block. An unspoken etiquette exists regarding the
use of public toilets. It includes the following:

- If you soil the toilet bowl, brush it clean before
  leaving the cubicle.
- Flush the toilet after every use – a public convenience is not the place to save water.
- If toilet paper drops to the floor, pick it up using
  another piece of paper.
- Leave the toilet as clean and tidy as possible.

Composting toilets

In dry locations, public composting toilets offer
travellers convenient facilities. Composting toilets
are excellent, especially when water is scarce or
unavailable. The pan is positioned directly over
the composting chamber. Within the chamber,
microorganisms turn waste into compost. A solar-
powered fan is sometimes used to remove odour.
To provide healthy, clean toilet facilities for

travellers, these systems are being installed all over the country. Always take note of instructions and leave the building clean and tidy for other travellers.

Camp toilets

Basic camp toilet

If there are no public toilets, select a private place at least 100 metres from any water course, campsite or track. Use a trowel or spade to remove any ground cover and then dig a hole at least 15 centimetres deep. Do what you need to do and if there are no fire restrictions in place, burn the toilet paper and re-fill the hole with soil. Replace ground cover and wash your hands. Alternatively, place soiled toilet paper into a biodegradable plastic bag, seal it up and dispose of it in the next town's tip or rubbish bin. In alpine regions, try to restrict your toileting to below the tree line.

If camping on or beside a deserted beach, dig a toilet hole near low watermark for faster decomposition of faeces. Don't bury your toilet paper, simply place it in a biodegradable plastic bag to be disposed of later. If you're using a campfire, burn the toilet paper in the fire – after you've finished cooking. If you're not using a fire, dispose of the toilet paper in a rubbish bin or tip in the next town. Nothing looks worse than toilet paper littering a beautiful place. Even worse is unburied faeces.

In environmentally sensitive areas such as alpine regions you are required to carry out all your toilet waste in well-sealed, heavy-duty biodegradable bags.

From Robin's travels

On a recent trip away, we camped at a free campsite, on the shores of a lake. The toilet block was located only metres from the lake's edge, with the septic system banked up and smelly to its rear. Clearly the toilet facilities were unsustainable; in fact, they were a health hazard.

Pit toilets

Set up within the privacy of a small tent, a Thunderbox toilet (positioned over a deep hole dug in either earth or sand) is the traditional toilet used when bush camping. Place a small trowel beside the toilet. After use, place a covering of sand or earth over faeces and toilet paper.

For convenience, a plastic container can be placed indoors for emergency use at night. Used for urine only, the container should be emptied first thing in the morning. Rinse clean using white vinegar and water, then leave to drain.

Portable chemical toilets

Portable chemical toilets are considered more environment-friendly than earth pit toilets (a deep hole dug in earth or sand, with a seat on top), which must be backfilled before leaving an area. Marketed as 'Porta-pottis', these systems are convenient for camping, caravanning and boating, and are becoming increasingly popular. They are easy to clean and relatively smell-free when used sensibly. Comfortable and fully portable, they flush toilet waste into a waste tank in a hygienic manner. They reduce the chance of introducing harmful microorganisms into creeks and rivers.

Dump-point locations

Major caravan parks usually provide a facility where you can dump toilet waste (black water). Those caravan parks with septic systems prefer biodegradable or eco-friendly toilet waste – rather than waste from chemical toilets that may contain harsh chemicals such as formaldehyde. It's always important to leave the facility clean. Official dump-point locations are found throughout Australia and are listed in various travel books. For example, Philip Procter's book *Camps Australia Wide* contains a list of official dump-point locations. If you're unable to dispose of toilet waste at a dump-point location, bury it at least 15 centimetres deep and 100 metres from any waterway, track, road or other campsite.

RUBBISH

'Carry in, carry out'

The policy of 'Carry in, carry out' is increasingly common in areas of state forest and other public land. If there are no rubbish bins provided, you must take away all your rubbish. It's not a good idea to bury food scraps, as native and feral animals will dig them up and scatter them far and wide, as well as eat foods unsuitable for their needs. Make it your aim to leave the campsite cleaner than you found it. Tidying up someone else's rubbish will give authorities reason to keep the site open for use. Tidiness is catching.

When camping for a weekend, take your rubbish home for recycling and disposal. Even small items like cigarette butts, citrus peel and silver paper need to be carried out. They don't decompose readily and look unsightly. Always carry strong rubbish bags in which to store rubbish. A litter bag kept in your car is always useful for small items such as tissues, drinking straws and food wrappers.

Recycle, reduce, reuse

The concept of recycling, reducing and reusing is important both at home and while on holidays. Unfortunately some people neglect this practice when travelling. Once you establish a system, though, it's easy to recycle, reduce and reuse.

Recycling waste

Paper, glass, steel, aluminium and some plastics can go into recycling bins. Broken glass, lying on the ground or in beach sand, presents a risk to animals and humans alike. Country towns usually recycle, so it's well worth your while to sort and separate your rubbish for recycling.

Some island communities set a high standard. Lord Howe Island has a sophisticated rubbish disposal system in place. All paper and green waste is chopped up in a shredder then spread as mulch. Netting is placed over the mulch to prevent it being blown away, or scratched away by birds. This netting is reused many times. All food scraps are composted in a central location. Flattened metal, glass and plastic waste is taken back to the mainland.

Reducing waste

In an effort to reduce waste, many country towns have a bulk food store. Travellers with the environment in mind shop at bulk food stores wherever possible. Purchasing local produce (such as fruit, vegetables, eggs and honey) from roadside stops and buying meat from country butcher shops also helps reduce the amount of packaging, and saves you money. When you pack to go away, avoid taking excess wrappings and containers – especially plastic and polystyrene packaging that's not biodegradable or able to be recycled.

Reuse

Reusing doesn't suggest that you're mean or living on the breadline, but rather that you practice the sensible reuse of certain products. This may include containers of all sorts, plastic bags (which inevitably sneak into your life), padded postage bags, wrapping paper (a quick iron will make it look as good as new), cardboard boxes, paper bags, and the backs of envelopes (for making lists and so forth).

By exercising your creative talents you'll be able to think of a lot more things you can reuse. For instance, a firewood box in rented accommodation may yield ancient treasures like ten- to 15-year-old newspapers and magazines. Reusing reading material can be great fun.

An opportunity for fellow travellers to share reading material exists at many caravan parks. At Robinvale, for instance, the laundry contains a small library of information books, magazines and novels. Reusing reading material is a sensible green concept. Travellers often pass on local maps and tourist information booklets as well. This is another good way of reusing printed material. Moreover, sharing always results in conversation that leads to further word-of-mouth exchanges of recommended places to camp, things to see and avoid and so forth. And before you know it, your washing will be done and pegged on the line in the sun.

From Robin's travels

Pouring vegetable oil into an empty but well-washed detergent bottle seemed like a good idea. Of course, it'd be easy to tell the difference! Placing the detergent and cooking oil bottles in the same cupboard also seemed sensible, especially in view of their identical height and bulk. After all, our VW Kombi van didn't have much space. After two weeks of outback travelling with no red meat at all, we spent an afternoon looking forward hungrily to the rissoles we planned to eat that night. While Doug lit the campfire, I groped in the half-light for the oil in which to cook the rissoles. Five minutes into the cooking cycle I lifted the lid. A pan full of bubbles and froth alerted me to my mistake. Canned tuna was on the menu, yet again.

My lesson was threefold. In future I would:

• carefully label reused containers
• always store food stuffs away from cleaning products
• never place food items in cleaning-product containers.

This way the green practice of reusing containers was both safe and practical.

Burning rubbish

Some travellers mistakenly believe that by burning all burnable rubbish they are helping the environment. True, they're reducing the amount that ends up at the tip. However, burning plastics – present in most milk and juice cartons and food wrappings – pollutes the air breathed by fellow campers, as well as every other living thing in the surrounding environment.

If you have a campfire, it's certainly okay to burn your paper waste. However, check first for hidden plastic before throwing anything into the flames. Toilet paper, kitchen paper and tissues are suitable for burning. Disposable nappies are not.

Adopting the old practice of 'burn, bash and bury' to dispose of cans is no longer acceptable. Carrying out your rubbish and recycling whenever possible is the modern approach.

Rubbish and aquatic environments

Anglers have a special responsibility. Empty bottles, six-pack rings, cans and wrappers spoil many fishing environments. Tangled and discarded fishing lines, fish hooks and plastic bait containers are just some of the residue left behind by careless anglers. This fishing litter is responsible for unimaginable pain and distress. Water birds, turtles, sea mammals and fish (as well as land animals and birds) become tangled in the litter or swallow it, leading to slow, painful deaths in the hundreds of thousands.

Plastic bags, cast aside and floating in the water, confuse birds like pelicans and mammals such as dolphins into thinking the floating bag is a tasty jellyfish. They eat the plastic bag and then die a lingering death.

Water pollution destroys the habitat of all water creatures, including fish. Therefore, when anglers fish from a boat, it's important to avoid any oil and petrol leaks and spills, however small.

Camping along the shore also needs to be done thoughtfully. Soaps and detergents pollute lakes and streams – even in small quantities – upsetting the chemical balance of the water. By doing your washing in basins or buckets (rather than in a lake or stream) and then pouring the waste water into a grease pit or scattering it widely, you'll be helping keep the fish's habitat intact. A healthy environment means more fish to catch!

GREY WATER

'Grey water' is the term used to describe all waste water, whether it's from a caravan park or bush camp. Remember that grey water will contain whatever soaps, detergents, disinfectants and cleaners you use while camping. This is why you should choose biodegradable washing powder or liquid, labelled low phosphate or phosphate-free. Eucalyptus oil and washing soda need to be used sparingly, even though they are green products.

Always wait until you have a full load before using an amenity block washing machine. That way you'll be reducing your impact on the environment.

Disposal

Untreated grey water may contain harmful bacteria, viruses and pollutants. These could pose health (to humans, pets and other animals) and environmental risks. The following hints will guide you towards the safe disposal of waste water:

• When washing clothes, utensils or yourself, do it 100 metres or more away from creeks, rivers, lakes or storm-water drains. Never put soap, detergents, food scraps or anything else into any water course. Use as little soap, detergent, shampoo and oil as possible. When washing dishes, sand and scourers are often sufficient.

- Do not allow pets to drink grey water.
- As washing water can clog the soil and damage vegetation, don't throw grey water onto the ground or over plant foliage. Instead, dig a grease pit: a hole about 45 centimetres square by 30 centimetres deep. Fill the hole with dry grass and leaves, then pour washing water slowly through the leaves and grass, which will collect any grease, soap residues or food particles in the water. Avoid saturating the soil or creating 'ponds'. Food particles attract flies, rats and mice, so take care that bits of food don't accumulate. Every day or so (depending on use), burn or bury the leaves and grass, and replace with fresh material. If the area is sandy, the sand will act as an additional filter to remove bacteria and nutrients. You may also choose to have your caravan sink draining directly into a specially prepared natural grease trap of this sort.
- If you cannot dig a grease pit, scatter your used water so that it filters through the soil, rather than harms vegetation.
- Never allow your waste water to enter a neighbour's camp.

Chapter 8
Travelling with Pets

Taking pets with you on holiday provides companionship and can enrich your overall experience. For example, travelling with dogs automatically gives you an affinity or point of introduction with every other person on the road who loves dogs. Animals are often the trigger through which friendships are born.

Identification

Identification is crucial. Your dog or cat's collar must carry an ID disc, a council registration disc and a contact telephone number. A microchip is even more essential, though, as it provides a permanent form of identification if your pet's collar and tags are missing or unreadable.

Pet ID can be tricky when you're travelling. However, a special travelling disc can be made up with your animal's name, your car's registration and a mobile phone number. For example, *Gus, travelling in White Toyota VIC DWB 687. Mobile phone 0426 774 423.*

Vaccinations and general health

Before setting out on your journey, make sure that your dog or cat's yearly vaccination is up-to-date, including heartworm treatment. Carry a current copy of the vaccination certificate in your glove box, in case your pet needs to be boarded as a result of an emergency of some kind, or in the event of you wanting to spend time in a national park. Carry pet insurance paperwork if applicable.

Keeping your dog or cat clean, well groomed and without fleas makes you and your pet more popular with the general public. Internal parasite treatment must also be considered prior to your departure.

All travelling dogs and cats need to be de-sexed.

Regulations

National parks do not allow pets, no matter how well mannered and controlled they are. Ask locally and you'll find a boarding kennel or cattery where your pet can be cared for while you enjoy the national park. Alternatively, you will find that the surrounding countryside offers scenery very similar to that within the national park.

When using caravan parks and other forms of accommodation, it's essential to ring the accommodation beforehand and ask if your pet will be welcome. Attitudes and owners can and do change.

DOGS

More and more caravan parks and other forms of accommodation are becoming dog-friendly.

The reason is probably a combination of the following:

- More and more people are travelling with their dogs.
- These enterprises recognise that by excluding this growing sector of the market, they're adversely affecting their profit margin.
- When dogs are in residence – within a caravan park, for instance – the overall security is improved.
- Travellers without dogs, and even those who don't like dogs, recognise the improved security, so choose to stay in dog-friendly forms of accommodation.

Bush camping opens up the possibility of enjoyable camping with all the family, including pets. There are hundreds of attractive bush camping sites all over Australia and several books list details of these locations. See the Further Reading section for details.

Did you know that you are 17 times more likely to talk to people when walking your dog?

The best breeds for travel

Any well-mannered dog that has been well socialised as a puppy, as well as obedience trained, will usually adapt easily to a travelling routine. An unruly, disobedient dog is never a welcome dog. In addition, your dog may make you unpopular and stressed – not a good holiday feel. The dog breeds that travel best usually have short coats. Long-coated dogs are difficult to manage in areas where scrub ticks, grass seeds and burrs occur. They also suffer more with over-heating in hot places. An agile dog is an advantage as it's able to leap in and out of vehicles. A good watch dog will guard your camp without being aggressive. Some good breeds include: boxers, dalmatians, German shepherds, German short-haired pointers, labrador retrievers, weimaraners, whippets, smooth-coated chihuahuas, fox terriers and Jack Russell terriers. Of course, there are many other breeds and crossbreeds that make delightful travelling companions.

From Robin's travels

At 15 years of age, I travelled with my family on a camping holiday to the Snowy Mountains of New South Wales. Caesar, my family's cream labrador dog, came as well, along with a Canadian canoe tied onto the roof-rack. A close bond existed between Caesar and myself, so much so that when I took the canoe out onto a lake, it almost ended in disaster. The moment I became aware that I would not make it back to shore easily, Caesar plunged into the icy water and swam frantically to 'save' me. He ended up nearly capsizing the canoe. We both made it back to the fringe of the lake, but not before I wondered seriously about the value of a dog on a canoeing holiday. On the other hand, if I had fallen in – his fault or otherwise – Caesar would have let me hang onto his otter-like tail and towed me to safety.

Travelling in a vehicle

Although dogs can travel on trains, ferries and by air, the most common way of travelling with dogs is in the family car.

An unrestrained dog that misbehaves within a vehicle causes driver distraction that may end in tragedy – especially in the event of sudden braking. A loose dog in a car is like a missile. When trained to travel in a car as a puppy, most dogs travel very well. Every dog is different, though, as is every travel set-up. You may need to experiment to find the best way for you to travel with your dog. Some of the options include:

- A very small dog will probably travel best in a small crate containing its favourite bed. Secure the carry box with a car seatbelt.
- Most dogs travel well in the back of a station wagon, separated from passengers by a properly mounted safety grill fixed between the back seat and the rear of the vehicle.
- Your dog can be trained to sit quietly on the back seat on a favourite blanket, wearing a harness secured to a back seatbelt.
- Many dogs travel well in the back of a utility or dual cab, with a fibreglass canopy to protect them from sun, wind and rain. A well-padded bed with a blanket on top will keep your dog comfortable. Elderly dogs with weakened or arthritic hips may need a portable ramp, in order to get in and out of a vehicle of this type.
- A dog trailer can be an excellent way to transport and house your dogs.

What to pack

- Pack a lead, a car harness (if required), a water bowl, a plastic scoop, plenty of biodegradable plastic bags and a container of water for rinsing your hands, and for offering the dog a drink.

- Gather together your dog's favourite bed, some towels, nail clippers and brush, as well as a few favourite toys.
- Pack plenty of your dog's normal food, as well as a feeding bowl. Many people find that a good quality dry food is best. Don't experiment with different types of food, though. Use a brand and type that your dog is familiar with.
- Take a first aid kit that includes your vet's phone number, any specific medication your dog may be on, bandages, cotton wool, saline solution, burn cream, Styptic powder (to stop the bleeding of a torn toenail), activated charcoal tablets (in case of accidental poisoning or bloat), tweezers, scissors, eye dropper (to force medicine or fluids into the mouth of an animal), antibiotic ointment, insect-bite lotion, eye rinsing solution, sunscreen (dogs with white skin around their eyes and nose are particularly prone to skin cancer) and antiseptic wipes.
- Pack a muzzle in case baits such as 1080 poison have been laid in the area, or your dog is in severe pain and likely to bite.

Departure

On the morning of departure:

- Feed only lightly, if at all, as most dogs travel better on a relatively empty stomach.
- Offer water.
- Take your dog to an area in the garden where toileting normally takes place, then give the command, 'Have a toilet.' This is an exceptionally useful command to teach any dog, as it makes toileting while travelling much easier.
- Double-check ID on your dog's collar and harness.

Dog etiquette

Dog etiquette involves a sensible blend of the following:

- Always pick up your dog's droppings.
- Place any dog hair in a biodegradable plastic bag and dispose of in a rubbish bin.
- Do not take dogs into national parks.
- Do not allow your dog into the amenities block of a caravan park. Dog paw prints are unattractive to most people.
- Dogs can kill native wildlife as well as sheep, goats and calves. Unless your dog has been trained never to chase sheep, cattle and native animals such as kangaroos, place a lead on your dog whenever in farmland or when native bush animals are around.
- Within caravan parks dogs must be walked on leads at all times, trained not to bark and have all their droppings picked up. You must never give anyone reason to complain about your dog.
- Never let your dog scavenge or wander away towards other campers.
- Keep your dog on a lead if you're anywhere near a road.
- Do not allow your dog to enter toilet blocks, school grounds, shopping centres or wander around children's playground equipment.
- Don't leave your dog tied up while you're out enjoying the sights. When you travel with a dog you have an extra responsibility. If you don't want this responsibility, leave your dog in a boarding kennel.
- During the summer months, popular beaches frequently have *No Dog* signs.

- If your dog is required to be on a leash at all times, make sure that it is.
- Barking dogs are never welcome.

Dog droppings and E. coli

Never let your dog do its droppings on pavements or other people's lawns or gardens. This is a legal requirement as well as a social responsibility. Hygiene is an important issue. There are a number of ingenious dog leads on the market, with built-in pooper-scoopers and biodegradable bags attached, but all you really need is a supply of biodegradable plastic bags. By using the bag as a glove, then turning it inside out and tying off the contents, you can pick up after your dog with relative ease. Place the bag (tightly sealed) in a rubbish bin. Some progressive councils provide biodegradable bags and bins in public dog-walking areas. Biodegradable bags can be obtained by phoning the DogTidy Company on 03 9753 6203.

By following this procedure you'll be safeguarding and improving water quality, especially after heavy rain when E. coli is flushed from stormwater drains into rivers and bay beaches. Faeces from any animal (including humans) contains E. coli, along with many other potentially harmful microorganisms.

On-the-road dog care

Safe enjoyable travel

- Travel sickness can be caused by anxiety, too much excitement or too much food prior to departure. Generally speaking, though, by gradually increasing the length of car trips, you can overcome this problem. If not, ask your vet for medication to help your dog overcome motion sickness.

- If you need to tie up your dog, first attach a loop around a post or tree, then attach your dog's lead to the loop. Check the safety of the position in regard to vehicles, ants and other possible hazards. Always provide shade and water, even in winter.
- In paralysis tick country (coastal region of eastern and northern Australia), carry out a daily search of your dog, especially within its ears and nostrils, in the lip folds, under the legs and neck, between its toes and in any skin folds. Using your fingers, feel for any small unusual bump.
- Check your dog's eyes and ears for grass seeds and irritating dust.
- Feel between your dog's toes for grass seeds or burrs.
- Keep your eyes open for snakes, and never let your dog bark at or chase a snake.
- Never let your dog run towards or around people who are fishing. A fish hook embedded in your dog's mouth is a distressing medical emergency.
- Put on a muzzle if there is any suggestion that baits such as 1080 poison have been laid.

Heat stress

Did you know that the inside of a car can reach 40 °C on a 25 °C day in only three and a half minutes, even with the windows open a little? You must never leave your dog locked in a car on a warm day, even when parked in the shade with windows slightly down. Death can occur within six minutes. Dogs with long hair and short or flat faces (for example boxers, Pekingese or pugs) are at increased risk. A responsible adult must stay with the car and dog, making sure that shade, water and good through-ventilation are provided. Alternatively, take your pet with you on a lead.

Two exceptions to the above rule occur when:

- your pet travels in a properly designed dog trailer
- your pet travels in the back of a utility or dual cab.

Both options rely on better-than-average roof insulation, a design that gives efficient through-ventilation and the provision of a bowl (which cannot be tipped over) containing water.

Panting, slobbering, whining, vomiting, diarrhoea and high body temperature all signal heat stress, and can lead to permanent brain damage or to coma and death.

If an animal is heat stressed, it's vital to cool down the body as quickly as possible – concentrating especially on the head and the back of the neck. Use cool water to lower the body temperature, but take care not to chill. Seek immediate veterinary attention.

Safe, clean places to rest

Along with creatures such as fleas, scorpions and ants, small fine-barbed burrs and grass seeds can be a real problem if they penetrate the soft skin between a dog's toes. In addition, your dog's coat will, of course, become soiled if it lies directly on dusty or sandy ground.

When camping, your dog needs a safe comfortable place to rest. This may be in the back of a utility, or wherever else your dog travels. A soft mattress covered with a heavy wool mat or blanket (easy to shake and air in the sunshine) can prove ideal for dogs in the back of a dual cab, station wagon or utility.

An alternative arrangement is a trampoline-type folding bed, located within the shade and shelter of an annex, and positioned close to your caravan door. This keeps your dog up off the ground. But make sure there's no ant nest or ant highway underneath! This type of bed comes in three sizes, and because it's easily folded into a remarkably light, compact tube, it's ideal as a camping dog bed. It has six feet with arms that fold inwards like an umbrella. A chain or rope will prevent your dog from wandering off.

Thick plastic mats that fit together like a jigsaw puzzle give your dog a clean, soft place to rest on the ground, free from dust, sand, burrs and grass seeds. Each measures approximately 60 centimetres square. Purchase the number required, depending on the size of your dog. In a matter of minutes, you have a quick lunchtime bed for your dog. These mats are also useful for humans – as doormats (the circular holes allow sand to fall through), exercise mats or sitting mats.

Fleas

Fleas thrive in dusty places. When travelling with a dog, it pays to keep your dog away from dusty public areas like park seats and tables, which may harbour fleas. Likewise, to prevent your dog digging a sleeping hollow in the dust beneath your vehicle while camped, it's a good idea to allow it free access to its travelling bed within the vehicle or caravan.

Hints to keep your pet's bedding and environment free of fleas include:

- A hard floor surface is the most practical surface for a caravan or motorhome. Mop using a solution made up of one cup of white vinegar to half a bucket of very hot water.
- Pet bedding needs to be washed regularly, using eucalyptus oil in the rinse water. Whenever possible, dry bedding in the sun for three hours. This will kill eggs and larvae.
- Any pet bedding used within the car should be washed regularly.

Safe, clean places to exercise

Frequent stops every couple of hours are essential when travelling long distances. These give you an opportunity to interact with your pet, stretch everyone's backs and legs, breathe fresh air, have a drink of cool, clean water and go to the toilet.

Due to their lack of burrs, thorns and prickles, dry creek beds are an excellent place to exercise dogs, as are outback airstrips and gravel roads. Similarly, almost every country town has a small public park and a large recreation reserve. Always carry some biodegradable plastic bags and a lead that you can clip on quickly if the need arises. For those dogs that pull badly on their lead, a Halti lead is the answer. This special lead offers an ingenious way of controlling your dog, without dislocating your shoulder or choking your dog.

Many dogs enjoy swimming and this can be good exercise. Allowing your dog to jump into unknown water is, however, a great risk in terms of hidden snags and water that's too shallow. Water that's too hot is another hazard to keep in mind. Some dogs have been badly scalded by plunging into a pool containing artesian or thermal spring water that's too hot. Always test the temperature of the water before allowing your dog to swim. Don't allow your dog to swim if other people are in the pool.

> There are times when it's permissible to disobey a sign. For instance, if you have a dog with you and a health crisis means that you need to stop in a place that says *No Dogs*, then stop regardless. It's unsafe to proceed in a situation such as this, especially if it's nearly dark. You get no medals for obeying a sign that is clearly not appropriate in an emergency situation.

CATS

Any cat that's been trained to live indoors will adjust well to life in a reasonably large-sized caravan or motorhome. Siamese cats, along with Burmese and Tonkinese, tend to adapt more easily to life in a caravan than the typical moggy. They are also easier to train to a harness and lead. You can be sure that every cupboard and shelf within its space will be climbed on, as every object and place within the caravan offers an outlet for your cat's intense curiosity and sense of play. Leaping from bed to bed, burrowing beneath doonas, sleeping on its special lamb's wool bed, and playing with its many toys will keep your cat entertained, while the same time providing many amusing moments for you.

Envirocats

It's no longer okay to let your cat roam. An envirocat is a new idea to a lot of people – a modern idea with surprising benefits. As well as benefiting wildlife, cats themselves are protected from poisonous baits, snakes, the danger of being run over by cars, or infected with parasites such as fleas, ticks or intestinal worms. Since contagious diseases such as cat flu and AIDS are spread by feral and roaming cats, an envirocat is always a more healthy cat. Cats and wildlife must be kept apart. It's a myth that cats need to roam.

In order to give your cat a safe space out-of-doors, you may choose to carry a specially designed cat enclosure that clips onto an outside caravan wall. This enclosure can be easily put up, dismantled and folded for transport. A small adventure playground, with access to your van via a cat flap or window, enables your cat to sleep in the sun, bird watch, climb and scratch its claws on a piece of rough bark. Envirocats need exercise, plenty of toys, playtime and affection – as well as a safe warm place to sleep. A travelling envirocat experiences every new place as an adventure. New birds to watch, new people, new everything!

Travelling in a vehicle

When trained to travel in the car as a kitten, most cats travel quite well. Given that every cat and travel set-up is different, however, you'll need to experiment to find the best way for you to travel with your cat.

- Most cats travel safely in a well-ventilated carry box, especially if their favourite blanket and toys are included. Secure the carry box with a car seatbelt. If your cat finds travel a scary experience, drape a towel or blanket over the carry box. Do make sure, though, that your cat has plenty of air to breathe and doesn't get hot.

- A well-adjusted cat may prefer to sit quietly in its favourite basket with its harness and lead attached and secured to a car seatbelt. These cats have learned (with practice) that being inside a car is safe and you can see all sorts of fascinating sights through the windows. They know from experience that complaining and leaping about while the car is in motion is totally forbidden and that this behaviour results in immediate confinement to their carry box. Always have a carry box handy in case of emergency. A distressed and non-restrained cat is a dangerous distraction, and in the event of sudden braking or accident, a potential missile.
- Some cats travel best when allowed to burrow beneath the doona on a bed within your caravan or motorhome – emerging sleepily on arrival at the next place.

What to pack

- Pack a litter tray, biodegradable cat litter, a plastic scoop, biodegradable plastic bags and a container of water for rinsing hands, and for offering your cat a drink. There are many sophisticated litter trays on the market, including automatic, self-cleaning trays, so your cat's toileting need be no problem at all.
- Pack your cat's favourite bed, nail clippers, a soft brush and a flea comb, as well as a collection of favourite toys.
- Pack plenty of your cat's normal food, as well as a feeding bowl. Don't experiment with different types of food, but use a brand and type your cat is familiar with. A good quality dry food is the easiest.

- Take a first aid kit that includes your vet's phone number, any specific medication your cat may be on, bandages, cotton wool, saline solution, tweezers, scissors, eye dropper (to force medicine or fluids into the mouth), antibiotic ointment, eye-rinsing solution, antiseptic wipes and Cat-lax.
- In the weeks prior to your departure, it's a good idea to give your cat a few practice runs at home in your stationary caravan or motorhome. This will help your cat feel more relaxed for those first few days on the road. In addition, the location of your cat's litter tray, food and water bowls and bed will be familiar.

Departure

- Feed only lightly, if at all, as most cats travel better on a relatively empty stomach.
- Offer water.
- Put your cat in its litter tray and leave undisturbed for five minutes or so.
- Double-check ID on cat's collar and harness.

On-the-road cat care

When travelling long distances, your cat may need to use its litter tray. Your cat will be willing to use its litter tray on the car floor if the vehicle is stationary and the cat really wants to go. Use one of the biodegradable plastic bags you have packed to scoop out the droppings, and then wash your hands. All cat hair, cat droppings and cat litter must be placed in sealed biodegradable plastic bags and disposed of in proper rubbish bins.

Offer regular cool drinks of water to your cat. Food also needs to be offered.

Your cat must be harness- and lead-trained. While travelling, always keep the harness and lead attached so that your cat cannot escape, get killed or lost as the result of a sudden unexpected fright or accident. A lead attached to your cat's normal collar is not secure enough, as all cat collars must have safety elastic inserts that allow the cat to slip the collar off in an emergency.

Heat stress

The inside of a car can reach 40 °C on a 25 °C day in only three and a half minutes, even with the windows open a little.

Never leave your cat locked up in the car on a warm day, as heat exhaustion, brain damage or death can occur within minutes. Instead, have someone stay with the car and cat, to make sure the cat remains comfortable. Alternatively, wearing its harness and lead, let the cat share your picnic lunch.

Most caravans are well insulated, have a roof hatch and windows that can be opened to allow good through-ventilation. Consequently, a cat can be left safely in this type of caravan without fear of them over-heating, though it's necessary to provide cool water to drink. Of course, it helps if the caravan is parked in a shady place.

Panting, slobbering, vomiting, diarrhoea and high body temperature all signal heat stress and can lead to coma and death or permanent brain damage. Long-haired cats with flat faces are especially susceptible to heat stress.

From Robin's travels

'Three dogs and two cats? You're not really taking them, are you? They're banned from so many places:

national parks, lots of caravan parks and beaches. They'll be a nuisance and ruin it for you.'

Since we enjoy the company of our animal family, we took no notice of our well-meaning human family and friends. Dogs and cats make us laugh and smile, and they keep our blood pressure low through the simple act of stroking them. So why not take them on a five-and-a-half week caravan adventure?

With dogs as companions, we had more options regarding campsites. Good watchdogs provide a feeling of safety and security, even in isolated bush camps. When we exercised our dogs we exercised ourselves as well. We got the circulation going and absorbed the atmosphere of country towns and areas of bush as only walking lets you do.

True envirocats, Kim and Katie were raised to live indoors with their only outings being on a harness and lead, or within an enclosed garden run. The transition to life in a caravan suited them very well.

Promises of numerous campsites where dogs were permitted lured us to the shores of Lake Pamamaroo, located near Menindee, 110 kilometres southeast of Broken Hill. Our sandy beach, bordered by the lake on one side and desert scrub generously vegetated with burrs on the other, kept the dogs from wandering. They rested in the rear of the Toyota with regular swims and drinks from the lake. The cats were content anywhere, their prime requirements being a sunny position in which to laze the day away – and, of course, our bed at night!

Returning home, people asked, 'You won't be taking your dogs and cats again, will you?' Surely once would be more than enough, they suggested. However, we had a fabulous time and the animals increased our enjoyment. We'd met so many more interesting people, walked a lot more, camped more safely and learned all about burrs. Yes, we would take our animal companions away again!

BIRDS

A pet budgie, cockatiel or canary makes a lively and amusing travelling companion. Ideally, the bird needs to be finger-trained and able to live 'free' within your caravan or motorhome. Its small cage offers safety while travelling, a feeding and watering station and a place in which to sleep at night. Your bird will enjoy a bath every few days and a selection of interesting grass-seed heads that you'll find on your travels. Your bird will reward you with its cheerful chatter, engaging tricks and, in the case of a canary, its full-throated warbling song.

Heat stress

Birds are especially susceptible to heat stroke. In a matter of minutes, a bird locked in a stationary car on a warm day, even with the windows down a little, will suffer brain damage, then coma and death.

Most caravans are well insulated, have a roof hatch and windows that can be opened to allow good through-ventilation. Consequently, a caged bird can be left safely in this type of caravan without fear of over-heating, though it's necessary to provide cool water to drink. Of course, it helps if your caravan is parked in a shady place.

From Robin's travels

Friendship seems to flow more freely when you're on the road. So it was with Jeanette, Sam and their finger-tame budgie called Stormy. We were camped beside Lake Pamameroo in the Menindee Lake system; us with three dogs and two Siamese cats, them with a treasured pet budgie. When Stormy became critically ill we teamed together to save his life. Using an eye-dropper and some finely crushed charcoal from our animal first aid kit, Jeanette gently and patiently forced a water, glucose and charcoal mix into Stormy's beak. This blend was chosen because Jeanette thought Stormy may have eaten something poisonous from the sandy ground beside their caravan. With one wing trimmed, they sometimes let Stormy out the caravan and let him peck about on the ground. Fortunately, with our magic blend we saved his life, forming a close and lasting friendship along the way.

Chapter 9
Staying Healthy

Undoubtedly, it's easier, safer and more comfortable to stay at home! This chapter, however, helps explore the subject of staying healthy and comfortable, while at the same time enjoying the green travel experience. Before leaving, all the people in your group should have a medical check-up.

ON-THE-ROAD REMEDIES AND HEALTH TIPS

Many of these remedies serve a useful role in the treatment of minor complaints; however, it's important to see a doctor if symptoms persist or the complaint causes you any anxiety at all.

Let's have a look at some of these remedies and health tips, but first a word of warning: essential oils such as eucalyptus, tea-tree and citronella can be highly toxic when swallowed, even in very small amounts. Just one quarter of a teaspoon of eucalyptus oil can put a child in a coma, so keep all your essential oils out of the reach of children.

Essential oils can also cause skin irritation if used

in their pure concentrated form. No more than 25 percent is recommended for rubbing directly onto the skin, especially for those with sensitivities. Any contact with the eyes must also be avoided. Essential oils can be safely diluted with good quality olive oil or grapeseed oil. You can buy them at a pharmacy or health food store.

Aching feet
After a long walk in the bush, this is a common complaint. Add salt (or bicarb soda) to a basin of hot water then soak your feet for at least ten minutes.

Aching joints and muscles
Gently massage eucalyptus oil (diluted with olive oil or grapeseed oil) into affected areas until the skin feels warm. Repeat until the ache disappears. Eucalyptus oil soothes as it penetrates, and often gives quick relief. Finish by wrapping a hot towel around the painful area.

Adhesive bandages and sticking plaster (to remove)
To remove an adhesive bandage without pain, simply moisten it with eucalyptus oil and it will lift off easily.

Aloe vera (Aloe barbadenis)
When travelling for a lengthy period of time, a potted aloe vera plant is an excellent idea. It will thrive indoors on a sunny bench, or in partial shade, though it doesn't like frost or salt-laden winds.

Aloe vera is one of the oldest medicinal plants in history, and provides an antiseptic gel that heals as well as gives pain relief. It's sometimes known as the first aid plant. Simply cut off a fleshy leaf, then split open to expose the soothing gel. The gel can be used to treat burns and scalds, cold sores, scalp conditions, sun spots, insect bites, skin rashes, fungal infections and to reduce scarring.

Anaesthetic (while travelling)
Ice cubes are the most effective travel anaesthetic, numbing pain in five to ten minutes. Perhaps you've been collecting firewood and a painful splinter has embedded itself into your thumb? Or a fishing hook has penetrated your finger? If you apply an ice cube wrapped in a tea towel or handkerchief, the finger will be numb within ten minutes. Then you can remove the splinter or hook with your attention on the ice cube, rather than the probing needle!

Anaesthetic (on a picnic)
If you're by the seaside, juice from the pigface creeper may be used to ease the pain of sunburn, stings and abrasions. Simply squeeze the fleshy leaves to extract the soothing gel.

On a bush picnic, the juice from immature bracken stalks can be used in a similar way.

Antiseptic
A salt and water solution makes a very effective and inexpensive antiseptic. Simply dissolve half a teaspoon of salt in one cup of boiled water.

Athlete's foot
Athlete's foot can be picked up when walking barefoot in public shower blocks, so it's advisable always to wear rubber thongs or sandals. If you suffer from this fungal condition, apply apple cider vinegar to your feet three to four times daily. Continue for two weeks after the infection disappears.

Sprinkle bicarb soda inside runners to reduce unpleasant odour.

Burns
Care needs to be taken when cooking over a campfire, as burns are relatively common. Plunge the burnt area into cold water for at least ten minutes (or till burnt area returns to normal temperature). This will reduce the heat of the skin, prevent further damage to tissue, and reduce the pain.

If the skin is not broken, aloe vera gel will promote healing and give good pain relief. Any serious burn should be treated by a doctor.

Cramps
A hot water bottle or hot pack can be effective in relieving painful abdominal cramps.

Dehydration
Dehydration occurs as a direct result of not drinking enough fluids. Aim to drink sufficient water to avoid any sensation of thirst. Dehydration can be a problem when you spend all day on the road. To avoid this condition, carry a large water bottle that you can access easily and sip from regularly. It helps if the water is cool, so keep the bottle within an insulated cover. Whenever walking, it's a good idea to carry a bottle of water from which to sip regularly.

A simple test for dehydration involves taking a small piece of skin and gently pinching it into a fold. If dehydration is a problem, the skin will stay pinched for some time after being released – otherwise it will immediately spring back into place. This test works well for both humans and animals. Dark urine indicates serious dehydration. Urine should be almost clear. When dehydration occurs, the ability to think rationally and drive a vehicle becomes impaired. Death will result in extreme circumstances.

Vomiting and/or diarrhoea and fever can also be the cause of serious dehydration. Replacement fluids, minerals and salts must be given, in the correct concentrations. Glucose, honey or sugar will help the absorption of minerals and salts. Plain water is not sufficient. A mixture containing one teaspoon of salt and eight teaspoons of sugar (or glucose or honey) in one litre of boiled water is recommended, taken in small, frequent sips.

Eyes (red and sore)
While travelling, your eyes may become inflamed and sore due to glare, dust or eyestrain. Bathe eyes in a weak salt and boiled water solution.

Gastroenteritis (bacterial)
This serious condition requires medical supervision. However, as a first aid measure, drink plenty of boiled water to which manuka honey has been added. This will help clear up the infection, as well as replace vital nutrients and fluids. Think carefully about your water supply. Has it been contaminated in any way? See pages 55–6 for information about waterborne illnesses.

Heat exhaustion and heatstroke
Heat exhaustion and heatstroke can happen to anyone. It may follow vigorous exercise, such as energetic bushwalking, playing tennis, jogging on a very hot day, or exposure to extreme temperatures without being able to cool off. Babies, young children and the elderly are particularly vulnerable, especially if left in a hot car. Symptoms include headache, nausea, feeling faint and

extreme thirst, which, if left untreated, lead to confusion, collapse and death.

Lie the person down in a cool place, remove excess clothing, and sponge their face and neck with cool (not cold) water. A mixture containing one teaspoon of salt and eight teaspoons of sugar (or glucose or honey) in one litre of boiled water is recommended, to sip slowly. If patient cannot drink fluids or vomits them up, seek urgent medical attention.

Heat rash
Sitting in a hot vehicle for lengthy periods of time can cause a prickly red rash that's very irritating. Loose clothing made of natural fibres will help, along with sitting on a towel. To relieve the itch you can use fresh aloe vera gel, a bicarb soda and water paste, or a cornflour and water paste.

Honey (to treat infected cuts, burns, sores, skin ulcers, wounds, sore throats)
Before the days of antibiotics, an infected wound was simply filled with honey, bandaged and left for 24 hours. On removing the bandage the wound was frequently clean, with the bacteria dead and all foreign matter coming away on the bandage. New tissue grew rapidly after the 24-hour honey treatment. Nowadays it's recommended that you continue the honey treatment until the wound has completely healed.

Selected honeys (manuka in particular) can be safely used to treat wounds. Recent research shows that honey reduces inflammation, swelling, odour and pain. Infection clears rapidly and healing is accelerated because honey (unlike antiseptics) doesn't damage tissue. In fact, honey supplies vital nutrients to regrowing cells and minimises ugly scarring.

Bacteria are unable to survive when surrounded by pure raw, untreated honey. This natural yet complex substance causes the cells of bacteria to dehydrate and die. Honey also draws poison from stings, and infection and foreign matter from wounds. It's a potent antiseptic that soothes damaged tissue, while at the same time stimulating it to regrow.

When selecting a place to store your medicinal honey, choose a position in your caravan or motorhome that's dark and cool. Too much light or heat may destroy some of the honey's unique healing properties.

Every traveller's first aid kit should include manuka honey, for both internal (one teaspoon of honey is excellent for a sore throat) and external use. When buying your first aid honey, choose one that's labelled 'Active UMF manuka honey'. This indicates that it's been tested for its antibacterial activity: a rating of ten is equivalent in antiseptic potency to a 10 percent solution of phenol.

Hypothermia

Exposure to severe chilling can occur when bushwalking, especially in wet and windy conditions or after accidentally falling into cold water. Typically a person will be shivering and have cold, pale skin. Their behaviour may be irrational and their breathing slow and shallow. These symptoms suggest that the body temperature is dangerously low.

First aid treatment involves a *slow* warming of the person by dressing them in warm, dry clothes (paying particular attention to the head, neck and chest), resting in a warm place, and giving warm drinks and high-energy food like chocolate. After first aid treatment, see a doctor or go to the nearest hospital for a check-up.

Jammed finger
Unfamiliar doors and equipment, especially in your first few weeks away, make jammed fingers a common complaint. Plunge your finger into very cold water, and then apply an icepack to reduce swelling and bruising.

Nausea
Sip ginger beer, dry ginger ale or ginger tea.

Nettle sting or rash
While searching for firewood, you may accidentally touch stinging nettles. If direct skin contact is made, these small plants, with their triangular serrated leaves, can leave sensitive skin feeling decidedly uncomfortable. When you brush against a nettle leaf, tiny sharp hairs containing an irritant fluid penetrate your skin and then break off. It's these stinging hairs that cause skin irritation and a rash. You will find nettles growing on disturbed ground and on bush tracks in moist areas.

One remedy that you can apply immediately (if wearing gloves) is the juice of the nettle stem. This will help relieve the itch. A more effective antidote, though, comes from the dock plant. Simply squeeze the juice of the dock plant over the painful area of the sting. For other soothing remedies see pages 180–1.

Poultices (to draw out infection or the poison from a bite or sting)
You can make an old-fashioned poultice, which is

particularly effective in drawing out poison or infection, by using any of the following recipes. Three recipes have been included to ensure that at least one will contain ingredients that you have on hand in your travel set-up.

- Combine one tablespoon of flour and one teaspoon of honey (preferably manuka) with egg white to make a paste. Cover the infected area with the paste, then bandage. Replace the poultice twice daily.
- Wrap half a slice of bread without the crust in a piece of cloth, and then dip it into a bowl of boiling water. Remove from the water, twist the ends of the cloth to squeeze out excess water, and place directly onto the infected area. Make sure, however, that it's not going to burn the skin. Bandage firmly to contain the heat. Repeat every hour. This poultice is often remarkably effective.
- Grate one tablespoon of pure soap and soften with a little hot water. Take one teaspoon of white sugar and mix this with the soap to make a soft paste. Use to draw out infection.

Poultices also help ease swellings and sprains, and help draw out deeply embedded splinters. Here are a couple of extra hints:

- To prevent the poultice sticking to the skin, first apply a little vegetable oil.
- To retain the warmth of a hot poultice, place a hot water bottle on top of the poultice.

If the infection has remained the same or become worse within a two- to three-day period, consult a doctor.

Scaly patches (especially on back of hands, face and ears)

Break a piece off your aloe vera plant and then dab the thickened scaly skin with the fresh gel three times daily until it disappears. If this sun-damaged skin itches, bleeds, ulcerates or changes size, shape or colour, see a doctor straightaway.

Splinters

Apply olive oil to the splinter prior to removal for an easy, pain-free extraction. Sterilise the needle before probing beneath the skin. See Anaesthetic on page 142 and Poultices on pages 147–8 for more hints.

Stubbed toe

Hot sunny days while travelling often means bare feet and stubbed toes. If you end up with a toe that feels as if it's broken but isn't, place a cold pack on it straight away. Later, dissolve one table-spoon of washing soda ('Lectric' soda) in warm water and soak your toe for about ten minutes. This will help reduce swelling.

Sunburn

Permanent shade beneath an annex, umbrella or tent will protect you from the sun's harmful rays. When relaxing on the beach, sunscreen (30 plus) will provide protection against harmful UV rays; likewise a shady hat, sunglasses and protective clothing. Try to avoid exposure to the sun between 11 a.m. and 3 p.m., and be aware that ultraviolet rays penetrate cloud.

To help ease the pain of sunburn, apply a bicarb soda paste to the burned area, or cut a leaf from your aloe vera plant and use the gel to soothe the burnt tissue.

Thrush (vaginal)

If after consultation with a doctor you still suffer recurrent thrush, the following self-help remedies may be of use to you:

- Avoid wearing tight jeans and synthetic under-wear, especially while sitting in a vehicle for long periods of time.
- Wash your cotton underwear using pure soap, then rinse thoroughly and dry in the sun.
- Yoghurt (plain and natural) with a live aci-dophilus culture is invaluable both to eat and use directly on the vulva and in the vagina. It works because it helps keep the vagina fairly acidic. In addition, its live cultures play an important role in controlling the candida organ-ism. Use it twice a day for about a week, to soothe inflammation and restore the natural acidity of the vagina.

Alternatively, bathe the inflamed area as follows:

- One day vinegar (one part vinegar to three parts water).
- The next day saline (half a tablespoon of salt to three cups of water).

Follow this procedure for at least one week, using a basin and a cloth.

Tooth problems

Dab oil of cloves onto gums and massage gently with fingertips. Oil of cloves is both an antiseptic and an anti-inflammatory preparation that will fight bacteria and fungi, as well as help relieve the pain associated with dental procedures. To further ease the pain, fill a cotton or woollen sock with hot coarse salt and hold it against the jaw.

To plug a cavity (or the gap between teeth) for the

temporary relief of pain, use whole or crushed cloves, or cotton wool soaked in oil of cloves.

If you're unfortunate enough to knock out a front tooth, soak it in saliva or milk – never water – and see a dentist as soon as possible. If you are a long distance from help, wash the tooth but use only saliva or milk, then replace the tooth in its correct position, facing the right way. Seek urgent dental help.

Urinary infection or inflammation of the bladder
Unfortunately, urinary problems are relatively common amongst travellers. Lack of fluids can be a factor, so plan never to be thirsty. (See Dehydration on pages 143–4.)

As a precaution, cranberry juice (or cranberry tablets) taken daily make it difficult for bacteria to attach themselves to the bladder wall, where infection usually begins. Likewise, vitamin C is a useful preventative remedy. Taken daily, vitamin C alters the acid–alkaline balance of the urine, making it less comfortable for bacteria to survive in.

If you have an infection, drink large quantities of old-fashioned barley water to reduce inflammation and flush away harmful bacteria. To make barley water, simmer unrefined barley in water for about one hour. Let it cool, then strain, and add lemon juice to suit your taste.

PERSONAL HYGIENE

When you set off on your travels, it takes a week or so to relax into new routines – including those of personal hygiene. And with the ever-changing facilities offered at caravan parks, it's not easy (or even possible) to maintain some of your home routines. Then there'll be times when bush camping requires you to learn additional skills. With practice, you can have an all-over wash (including your hair), using just one small basin of warm water! Depending on the circumstances of your overnight stay, the following hints may be useful.

Deodorants

A light dusting of bicarb soda under the arms is a simple, safe and effective deodorant. Bicarb soda does not stop perspiration from occurring, it simply prevents the growth of odour-producing bacteria.

Full-strength apple cider vinegar (or herbal cider vinegar) is also effective as a deodorant for use under arms or on feet. As the vinegar dries, the vinegar odour completely disappears.

Body-rock is a natural deodorant that has been used for centuries in Asia and Europe. It's a naturally occurring mineral salt that is fragrance-free, pure and non-allergenic. To apply, simply moisten the rock and rub under the arms. The mineral salt prevents the growth of odour-producing bacteria.

Face

To cleanse

A mild pure soap is all that is needed.

To moisturise

After washing your face, apply sorbolene lotion (with 10 percent glycerine, as well as evening primrose oil and vitamin E) to your face. This is an inexpensive and excellent moisturising lotion that protects your skin against the elements.

Feet

When travelling in warm climates, walking barefooted (or wearing sandals or thongs) exposes your feet to unusual conditions. To gently reduce rough, hard skin or calluses, first soak your feet in a warm water and bicarb soda solution. As your feet soften, wet a pumice stone, and then rub it against a piece of soap. Now rub the pumice stone gently but firmly against the hard skin. Regular use of pumice stone, generously lubricated with soap, will keep your feet smooth and soft.

Dry, cracked skin on heels can be softened by the daily use of sorbolene lotion (with 10 percent glycerine, and added evening primrose oil and vitamin E). Massage the lotion into the skin to replace moisture and elasticity. Dust smelly shoes (and feet) with bicarb soda.

Hair
To condition
Dry, dusty conditions while travelling often cause hair to become dry and brittle. After shampooing, massage plain yoghurt (two teaspoons is plenty) through your hair, then rinse well. This gives a pleasing finish, especially to dry hair.

To rinse
After a normal shampoo, the addition of vinegar or lemon juice to the final rinse gives your hair extra lustre and removes all traces of shampoo. It also restores the correct balance between the acidity and alkalinity of your scalp. Vinegar is good for brunettes and lemon best for blondes. If your hair is extra oily, use lemon juice and increase the proportion of lemon juice to water.

To control dandruff
Combine half a cup of apple cider vinegar with half a cup of water. Apply to scalp before shampooing.

Anti-chlorine hair rinse

After swimming in chlorinated water, rinse your hair with a weak solution of bicarb soda and water.

To dry wash or rinse hair

If your hair needs a lift but there's neither time nor water for a normal shampoo:

- Sprinkle your hairbrush with bicarb soda, then brush your hair. This will remove dust and excess oil.
- Or, rinse with warm water to which you've added a little white vinegar.

Hands and elbows

The inside of an avocado skin rubbed into the hands or elbows combines abrasive and oil-giving qualities.

Lemon juice is very effective for removing stains. After washing, apply a few drops of olive oil to the palms and massage into the skin to soften and protect.

Lips

Cracked, dry lips are a common complaint when travelling. If regularly applied, vaseline lip therapy with sunscreen makes an excellent lip-covering to protect against cracking, drying and sunburn.

Nails

Stains made by handling blackened campfire equipment may be removed from nails (and hands) by soaking them in a solution of three cups of warm water to which one dessertspoon of lemon juice has been added.

Perfume

Consider leaving your perfume at home. By doing this you'll be able to smell the delicate aromas of the Australian bush, as well as the delectable salty tang of seaweed along the shore.

Teeth and gums

Dental hygiene must not be neglected while you travel. No one wants a visit to the dentist while on holidays. Salt mouthwashes help to disinfect your gums and keep them healthy. After dental work, a new tooth or a gum injury, use a solution of half a teaspoon of salt to one cup of boiled warm water. This antiseptic wash, used every hour or so, assists healing.

Bicarb soda paste is a safe and very effective cleaning agent for teeth. As well as its gentle cleaning and polishing action, bicarb soda deodorises the mouth and reduces decay by neutralising plaque. Simply use a little bicarb soda on a wet toothbrush. Oil of peppermint may be added for palatability, if desired.

An occasional application of lemon juice and bicarb soda will keep teeth shiny and white. Combine one teaspoon of bicarb soda with lemon juice to make a stiff paste. Apply with a brush, and then rinse mouth with cold water.

Dentures and plates can be cleaned using a toothbrush and bicarb soda. Overnight, place dentures and plates in a glass that contains equal parts vinegar and water.

Chapter 10
Staying Safe

In the same way that you can minimise damage to the environment through the choices you make and how you behave when you travel, so too can you minimise danger to yourself and your travelling companions. Being informed about the sorts of things you might encounter when you travel is the first step towards avoiding serious injury and staying out of danger.

Before leaving on your adventure, ensure that your vehicle (as well as anything you're towing) has been thoroughly checked and serviced. By doing this, you'll greatly increase the safety and enjoyment of your travel.

EMERGENCY PHONE NUMBERS

A list of emergency numbers needs to be kept in your first aid kit, as well as your vehicle's glove box. The following numbers offer the basis for your list.

- Ambulance, Fire, Police:
  - From a land-line, dial 000
  - From a mobile phone, dial 000. (If in an emergency you are not within mobile

phone range, key in 112 then press the *Yes* key to be connected to police, ambulance or CFA.)

- From a satellite phone, use the short code 13. A national operator will transfer you to an emergency service.
- Poisons Information: 13 11 26
- Marine Stingers: 08 8222 5116
- Royal Flying Doctor Service (emergency calls only)
  - New South Wales 08 8088 1188
  - Northern Territory 08 8952 1033
  - Queensland 07 4743 2802
  - South Australia 08 8642 2044
  - Western Australia 1800 625 800

For truly secure travel, a satellite telephone is the ultimate safeguard. See page 177 for a description of its capabilities.

Your location will need to be clearly identified. A map and grid references will be useful, along with the number of people injured and their condition.

TRAVEL ETIQUETTE

Practising respect for the people you meet and the environment through which you travel is the key to a safe journey. Travel etiquette involves awareness of the following points:

- Respect other travellers and road users.
- Consider local people, their property and stock.
- Never fiddle with water bores, tanks or windmills.
- Respect Aboriginal land and communities, and obey their signs. Obtain a permit before entering Aboriginal land. Keep in mind any rule

concerning the carrying or drinking of alcohol.
- Don't take photographs of Aboriginal people, sacred places or art without permission.
- Slow down on gravel roads (especially corrugated surfaces) to prevent your vehicle throwing up stones and damaging other travellers' windscreens.
- Clouds of dust, heavy rain or hail obscure vision. Slow down and turn on your headlights until you find a safe place to pull over. Let the dust settle or storm pass before continuing on your way.
- Pull over to the side of the road, when safe, to let faster-moving traffic pass.
- Road trains cannot drive on the soft edges of the road due to the risk of overturning. Therefore, if the bitumen is narrow, move off the road as far as possible.
- Wind resistance can be a problem, especially when a road train overtakes a car and caravan. Slow down (or preferably move off the road and stop) when you see a road train approaching.
- When driving a four-wheel-drive vehicle, remember it has a high centre of gravity and can be affected adversely by wind and large overtaking vehicles.
- Avoid travelling on wet unsealed roads, to prevent cutting up the surface.
- Don't camp on private property without asking permission.
- Leave gates as you find them.
- Don't be a nuisance to station people. Asking for petrol, gas or water is an unreasonable request. Expecting to be pulled out of a bog will probably cause irritation. So too will silly questions. Station people work hard and for long hours.

- Keep to roads and tracks.
- Avoid being noisy.
- Take care with fire. Don't let your campfire turn into a wildfire.
- Leave no litter.
- There will be times when you share the road with sheep, cattle, emus and kangaroos. These animals graze along the roadside, especially at dawn, dusk and during the night. They will be confused by your car's headlights, so avoid travelling between sunset and sunrise when possible, or take care.
- Use water wisely.
- When camping beside water, take care not to contaminate it.
- Think safety in all that you do: rushing causes mishaps.
- Care for native plants and animals, and their habitat.
- Respect the past. Leave old things as you find them.

DANGERS AND PRECAUTIONS

Bogged vehicle (sand or mud)

Travelling over unsealed road surfaces that are very wet causes unnecessary erosion and damage to the structure of the road. Maybe you can change your route, or wait until the road has dried out? Driving at a moderate to slow speed, and in keeping with weather conditions and terrain, will be better for your vehicle and the environment over which you're travelling. Heavy penalties may be imposed if you damage the road surface. Likewise, avoiding alpine meadows, wetlands, saltpans and sand dunes is wise environmental practice for four-wheel drives. These are sensitive areas that are best explored on foot.

When travelling through country that may be muddy, icy or slippery with snow, it's advisable to carry snow chains.

If your vehicle becomes bogged in sand, you can use floor mats to give support and traction. To avoid becoming bogged, drive slowly, without stopping. Keep to existing wheel tracks wherever possible. Keep the vehicle in a straight line. Any turns should be executed by turning the wheel rapidly in the direction you want, and then turning quickly back to straight ahead.

By reducing the pressure in your tyres you spread the vehicle's weight over a larger area. This improves your traction, prevents the wheels spinning and keeps erosion to a minimum. Remember to re-inflate the tyres once you're clear of the sand. Getting bogged almost always causes damage to the environment. Try not to go beyond the limits of your vehicle.

From Robin's travels

'What a bloody idiot,' declared a lone amateur photographer, poised on the dam bank to capture a sunset over mirror-clear water. 'He's bogged himself right in my shot.' I hesitated, then decided to own up before he said any more. 'Actually, he's my husband,' I admitted. 'And yes, we did make a foolish mistake.' The bloke spluttered an embarrassed apology and then, as he fiddled unnecessarily with his tripod, I steered the subject towards his photography. Soon he was up and away on his pet subject. After letting him go for five minutes or so, I broke into his spiel saying, 'Well, I'm supposed to be getting some help, so I'd better get going.' On the following morning we spent an hour or so filling in and smoothing over the damage our vehicle had caused on the deceptively firm-looking bank, below which lay a deep, soupy sludge. We should have tested it on foot before driving any further. It was a lesson we would never forget.

Breakdown

If your vehicle has been fully serviced and found to be mechanically sound prior to departure, breakdown is unlikely to occur. When planning to travel through remote country, however, leave details of your plans with someone trustworthy. In addition, check weather conditions and road distances before setting out on your journey.

If you break down, never leave your vehicle. A missing vehicle is much easier to locate than missing people. If you have an adequate supply of emergency water, fuel and food – as well as a first aid kit – you will be well equipped to survive until rescue services find you.

Apart from staying with your vehicle, the next most important advice is to remain calm. Make a cup of tea or coffee (or prepare a cool non-alcoholic drink), then sit down and plan a sensible strategy. Use maps and a compass to work out, as accurately as possible, your location. Important parts of your plan should include:

- food and water rationing. See also Water (finding in an emergency situation), on pages 173–4
- shelter from prevailing winds
- protection from heat and cold

- methods of attracting attention such as a smoky fire, mirror or magnifying glass to signal for help. Sunlight reflected from a small mirror can be seen over large distances, especially repetitive movement. To draw attention to your plight, spread white toilet paper around your camp, place mirrors on the roof of your vehicle, or burn green leaves to make smoke
- ways of counteracting anxiety such as playing games.

Ideally you'll have a Global Positioning System (GPS). This piece of equipment provides you with accurate figures relating to your position, height above sea level, speed and direction of travel. When travelling in remote areas, through a maze of unmade forest roads, on station tracks or bush-walking through isolated country, a GPS will be a worthwhile investment. Detailed maps (that include GPS coordinates) are essential to use in conjunction with a GPS receiver.

Emergency Position Indicating Radio Beacons (EPIRBs) are carried by some outback travellers for use in a life-threatening situation. When activated, an emergency radio signal is transmitted by the unit, received by a satellite, and then retransmitted back to a control centre. The signal pinpoints your position so that a rescue plan can be initiated.

Bulldust

Fine talc-like dust occurs in many outback areas. As surface dust it can cause damage to car engines. It also causes eye irritation. Even more dangerous is the fact that the surface of the road may look smooth, when in fact it's badly damaged. Therefore, drive slowly and carefully through country where bulldust is a problem – and wear protective glasses.

Bushfires
Whenever conditions are hot, dry and windy, it makes sense to plan your itinerary accordingly, especially in relation to the risk of bushfires. During the warmer months make sure you obtain daily fire danger information. Listen to local ABC radio for advice.

If caught in a vehicle during a bushfire, park in an area as free of vegetation as possible. Do not leave your car. Close all windows. Turn on your hazard lights and headlights. Tune your car radio to ABC local radio, for current information about bush-fires. Lie down as close to the floor as possible, and cover yourself with a coat or woollen blanket, to protect from radiant heat. Drink water to avoid dehydration. Stay in this position until the fire front has passed.

Bushwalkers need to take extra care, and should remember places they pass where they could seek shelter from a potential fire. This includes places such as a river or wet gully, areas of erosion where there's not much vegetation, a hut, a con-crete bridge, a road cutting, a gravel pit, or an area that's been recently burned. Never try to out-run a fire. To avoid breathing in heated air and smoke, keep as close to the ground as possible. Remember that radiant heat kills, so wear cloth-ing that protects all exposed skin – wool, close-weave cotton or oilskin garments are best. Synthetic clothing will burn and melt onto skin.

Bushwalking precautions
Always plan your walk with care, taking into account the time of year, an accurate weather

forecast, and the fitness level of the least-fit person of your group. Walking alone in remote areas is seldom a good idea. Even day walks involve careful preparation. Wear clothing and take equipment for the worst possible weather conditions. Take a compass and an up-to-date map, drinking water, emergency foods such as dried fruits, nuts or chocolate, matches, paper and pencil, a pocket-knife, a small first aid kit, sun protection (sunscreen, sunglasses and a wide-brimmed hat) and a waterproof jacket. Strong comfortable shoes with a good tread (sandals and thongs can be dangerous) are a must. Keep to tracks, don't cut corners, check your compass and map regularly, and keep your group together.

Before you set off on any adventurous trek, you must tell a responsible adult (preferably a ranger, the police or a family member or friend) of your proposed route and time of return. Avoid returning at dusk.

If lost, stop, sit down and discuss the situation calmly. Look for tracks and prominent features then retrace steps if necessary. If you can't retrace your steps, stop and make an emergency camp. Keeping dry and warm is a priority, so choose a place where rocks, hollow tree-trunks or trees with large canopies provide natural shelter and shade, if it's hot. Aim to keep out wind and cold using branches and brush as makeshift walls. Light a fire and keep it going. Ration your food and water. Staying in this emergency camp is usually the best strategy, though if you do move on, leave a note saying when you left and where you were headed. Leave a trail for rescuers to follow. Shout 'help' or 'cooee' regularly as you walk.

Creek crossings and flash floods

If possible, avoid driving through creeks because this washes polluting oils from the engine into the water.

When approaching water flooding across a road, stop to assess the road surface and the depth and flow of the water. Usually it's a wise policy to do a test wade prior to taking your vehicle through water. This way you'll test the depth of the water and the consistency of the track beneath the water before attempting a crossing. When crossing with your vehicle, drive slowly but at a constant speed, and keep to the centre with your wheels straight. Do not change gear midstream.

During the monsoon season in northern Australia, and any season in southern Australia, flash floods can close roads. Always obey road authority signs.

A wall of water (up to half a metre high) can surge down a dry creek bed, even in the middle of a drought. Likewise, water levels in some inland creeks can rise overnight by up to four metres.

Cyclones

Cyclone and Weather Information

- Tropical Cyclone Warnings (NT, QLD, WA): phone 1900 155 355
- National Weather Service: phone 1900 155 344
- Seasonal Outlook: phone 1900 155 348

The Wet in the Top End is a dramatic time of year. Extreme heat and humidity, thunder and light-ning, torrential rain and flooding, and rough seas and high tides occur during the monsoon season between November and April in northern Australia.

The violent and often destructive force of a cyclone poses considerable risk to travellers in the area. Caravans and tents are not built to withstand the force of cyclone-driven winds. For this reason, it's wise to avoid travelling in northern Australia during the Wet unless absolutely necessary. Cyclone Tracy, which flattened the city of Darwin on the Christmas Eve of 1974, is a cyclone all Australians will remember.

If travelling in a cyclone area, you'll need to listen to the weather reports on ABC local radio at least twice daily. If a cyclone is approaching your area (within 200 kilometres), head in the opposite direction. If you cannot leave, you may be required to chain down your equipment. Officials have the power to secure or carry away your gear during an official Cyclone Watch period if you do not cooperate in following safety procedures.

Dust storms

A severe dust storm makes safe travel impossible. Switch on your hazard lights and headlights and pull over into a rest area. Wait for the storm to pass and visibility to return.

From Robin's travels

A huge flock of corellas circled the lakeside caravan park, landing occasionally in the branches of dead trees while all the time screeching their agitation and distress. Distress about the gale-force winds, the air choked with red dust, and now a heavy rainstorm. I stepped outside and held out my arms. Large splashes of red landed on my skin. It was raining 'blood'.

As soon as the storm had passed, the corellas stopped their raucous screeching, soared and wheeled in the

turbulent air and flew away, protesting no more.
The lifeblood of the land, sacrificed by land-holders
desperate to sow crops after years of drought.
Understandable, perhaps, though as destructive as
the pollution, salting and overuse of the Murray
River's water. The dying of the land, the dying of
the mighty Murray.

Fatigue (beating fatigue while travelling)

When travelling long distances it's a sensible idea
to stop and rest at least every two hours. A brisk
walk helps blood circulation. Eating a light meal
or having a tea or coffee break helps as well.
Changing drivers on a regular basis allows you to
travel long distances with less fatigue. Further,
your comfort will be prolonged by wearing comfortable clothes, having sunglasses close by,
keeping the windscreen clean, having good back
support, not eating a heavy meal in the middle of
the day, drinking plenty of water and planning
your day so you arrive at your overnight stop
with plenty of time for relaxation.

Secondary sealed roads offer a valuable option to
major highways, especially in relation to fatigue
and quality of travel. On secondary roads there is
reduced traffic. This means that it's possible to
reduce your speed to 80 kilometres per hour without annoying other traffic on the road. When
towing a caravan, the reduction of speed from
100 kilometres per hour to 80 kilometres per hour
makes a big difference in terms of fatigue, safety
and fuel economy. Enjoying the surrounding
countryside is another benefit, along with the
opportunity to meet local people in small country
towns and wave to other travellers on the road.
Your travelling experience is more relaxed and

pleasurable on secondary roads. After all, who wants to set a record on the number of kilometres travelled in one day, two weeks or three months? Travelling isn't meant to be an endurance test.

Fire (within your caravan or motorhome, or from campfire)

Usually fitted with one access door only, most caravans and motorhomes pose a serious risk if a fire starts indoors. Therefore:

- Install two fire extinguishers. Place one by the exit and the other in the kitchen area.
- Place a fire blanket (to smother flames) in the kitchen area, close to the stove in case cooking oil ignites.
- Install a smoke alarm in the sleeping area.
- Do not lock the door from the inside and withdraw the key. In an emergency you may not be able to locate the key.
- Never block the gas vents in the door or floor. In the event of a gas leak, gas collects at the lowest point and will leave via the vent. Gas can kill.

Wool (resistant to fire), close-weave cotton or oilskin garments are the best fabrics to wear around a campfire. Most synthetics melt and stick to the skin if very hot, causing serious burns and complications. Sparks and tongues of fire make campfires potentially dangerous unless you're dressed correctly. Ideally you'll be wearing a woollen jacket.

Ocean rips

Like a strong river current, an ocean rip flows out to sea, usually from a surf beach. Even for strong swimmers it holds many dangers. Recognition of a rip involves looking for the following:

- seaweed floating out to sea
- water, away from the shore, that looks rippled, discoloured by sand or has foam on its surface.

If you're swimming between the flags on a patrolled beach and you find yourself in difficulty:

Relax

Raise your arm, and wait to be

Rescued.

If you're alone and in trouble, don't panic. Instead, swim across the rip, not against it – keeping parallel to the beach until you reach calm water. Most rips end just beyond the breakers and are no more than 30 metres wide. When you're clear of the rip, swim to the shore.

Swimming between the flags – at patrolled beaches – offers protection from shark attack, drowning, marine stingers, crocodiles and irresponsible people. Always read and obey warning signs, never swim alone, always supervise children, don't swim for half an hour after eating, and never mix alcohol or drugs with water sports.

Overtaking
Dust thrown up by an overtaking vehicle can reduce visibility to almost zero. Loose stones and dangerously soft edges pose extra dangers, especially if towing a caravan.

Road trains (which can be 50 metres in length and weigh 170 tonnes) need to be treated with great respect. When you see a road train approaching (from either direction), it's intelligent to pull right off the road and wait for the road train to pass and the dust to settle. Due to its immense weight, a road train cannot necessarily leave you half the road. It must stay central on the road and some bitumen strips are narrow.

When travelling along gravel roads, or narrow bitumen roads with loose stones on the edges, it's a good strategy to slow down whenever you see traffic coming towards you. By doing this, you'll reduce the likelihood of windscreen damage.

Permission (to enter private property or Aboriginal land)

Roads and tracks on private property should not be used without the land-holder's permission. Gates and grids across the road, or absence of fencing on one or both sides of the road, indicate the road may be on private property. Gates must be left as you find them.

Photography

There are places and situations where flashlight photography is inappropriate. The Phillip Island Nature Park's Penguin Parade is one of these. Penguins tumble out of the waves at dusk and march up the beach to their burrows in the dunes. During the breeding season their crops are full of food, ready to regurgitate into the mouths of hungry chicks. Flashlights startle the penguins resulting in regurgitation of food onto the sand. This means that the penguin chicks will die – it's as simple as that! Green tourism relies on people obeying signs and requests made by rangers.

Photographic equipment and film needs to be kept in a cool place. Heat will cause film to dis-colour and deteriorate. Carrying a spare camera battery makes sense.

In an attempt to protect their camera from dust and sand, some people place photographic gear within a plastic bag. Plastic, when exposed to heat or sunshine, may cause condensation to form within the bag, leading to damage of expensive equipment. A special camera bag, a cloth bag or a leather bag will breathe, as well as protect photographic equipment.

Quarantine

When travelling through major horticultural regions – such as those found along the Riverland area of South Australia, northern Victoria and southern New South Wales – it's important to be aware of quarantine restrictions. The ban includes all fresh fruit as well as tomatoes, capsicums and avocados. To protect these important horticultural areas, worth billions of dollars annually, the ban is policed by random road checks and on-the-spot fines. At fruit-fly exclusion zones – these are well signposted – you must stop if you're carrying fruit and either eat it or dispose of it in the bins provided.

Western Australia has strict quarantine regulations restricting the movement of fresh fruit and vegetables, as well as honey, into the state.

From Robin's travels

It's not smart to smuggle fruit and vegetables into areas such as Mildura, one of the most important fruit and vegetable bowls of Australia. While stopped at a quarantine drop-off point, with its signage and fruit bins, several other cars and caravans pulled in as well. Then we all began the serious business of eating as much of our fruit as we wanted, discarding the remainder into the bins. What a great opportunity to chat with fellow travellers: two couples from Tasmania, and a man who travels Australia pruning vines and picking whatever fruit is in season.

River red gums

Avoid camping under river red gums. Their branches are quite likely to break off without warning, in high winds or very hot calm conditions.

River safety

Camping beside a river holds a magic all of its own. There are, however, hidden dangers such as river red gums dropping branches, cliffs that are undercut and heavily eroded, snakes seeking a drink of water, swings attached to the rotten branches of trees, and within the river itself, snags, strong currents, icy water and deep holes.

Rock fishing

Amateur rock fishing by travellers can end in disaster, unless certain safeguards are kept in mind. Things to study beforehand include the size of the swell, as well as changing tides and weather conditions. By watching the ocean for about 20 minutes, you'll get an idea of the size of the waves (including the occasional king wave), and be able to plan a possible escape route if you are swept over the edge. Asking locals for advice, fishing with other anglers and wearing sensible shoes all help make your fishing expedition a happy and safe experience, regardless of whether or not you catch a fish.

Wandering stock and wildlife

Appetising green grass and weeds grow along the edges of roads, watered by run-off dew. For this reason, sheep, cattle and wildlife often graze these areas, especially on unfenced outback roads. As you travel through the countryside using secondary roads, slow down and make sure you scan both sides of the road, especially at dusk, during the night and at first light, when animals such as emus and kangaroos are most active.

If you're forced to travel at night, reduce speed so that if a kangaroo leaps out unexpectedly, you have a chance of stopping in time to avoid a collision. A kangaroo, blinded by the headlights of an oncoming vehicle, and panicked by the engine noise, tends to leap the wrong way and move straight into the path of a fast-moving car. A collision with a bounding kangaroo can be fatal for both the animal and the occupants of the vehicle – especially if a large buck kangaroo comes through the windscreen. In terms of vehicle damage, it makes a huge difference if you hit a kangaroo at 100 kilometres per hour compared to 80 kilometres per hour. At 60 kilometres per hour it's usually possible to miss the kangaroo altogether.

To prevent being blinded by an oncoming vehicle, look slightly to the left of the road.

Water (finding, in an emergency situation)
Sometimes water can be found by digging just a few metres beneath the surface. In a dry sandy creek bed, for instance, a fresh water soak may be located at the lowest point of an outside bend.

If by the sea or a dry inland lake, dig just a few metres above the high water mark and you may strike brackish water that will be much better than nothing. Alternatively, look for reeds growing in the sand dunes behind the beach and dig, or dig at the lowest point between two sand dunes. Coastal cliffs usually have small soaks and springs, recognisable by the growth of ferns or moss.

In desert areas, water can be found in the roots of many desert plants, the flesh of succulent desert plants and morning dew. Watching birds can lead people to water, especially at sunset.

If your water reserve is low, constructing a solar still can be a lifesaver. Choose a place in direct sunshine, and dig a hole about one metre wide by half a metre deep. In the centre of the hole, place a container to collect water. Now gather together green leaves, fleshy plants such as succulents and roots, and pack them around the container. Don't cover the container. Cover the hole with plastic sheeting and seal the edges with soil and rocks. Place a stone or some soil in the centre, directly above the container, so that the plastic sags.

The sun's heat will evaporate the moisture from the green leaves, roots and surrounding soil. This moisture will condense on the underside of the plastic, run to the centre and drip into the container. To drink the water, you can suck it through a straw or a plastic tube. About two litres a day can be collected this way, so you'll need to construct two or more solar stills to satisfy the needs of your travelling party. Every day you will need to replace the plants.

Another idea is to enclose leafy branches or shrubs within large plastic bags. Water will collect within the bags. Drain daily, then place the bags on fresh branches.

COMMUNICATIONS

Knowing that you can have confidence in the reliability of your communication systems goes a long way towards peace of mind. For this reason, it's crucial that you consider the various options available to travellers, and then make your decision in relation to the age group and the general health of your participants.

There are many options from which to select when considering a form of communication that will keep you in touch with family and friends, as well as emergency services. (For details about emergency services see pages 156–7.)

Email

Email provides an effective, inexpensive way to keep in touch with family and friends. You can use internet centres and cafés, public libraries or tourist information centres or your own computer.

HF and UHF radio

HF radio, with up to 500 channels, has a close connection to the Royal Flying Doctor Service (RFDS), and is a vital service for people travelling or living in the outback. HF radio is for use over long distances and requires a special aerial.

UHF radio is for use over short distances (usually less than 40 kilometres), is easy to use and inexpensive to purchase. It's a fun way of talking and gathering information from other travellers and truckies along your route. Its major use, therefore, is as a form of communication between vehicles. In an emergency, UHF radio would not be the communication of choice.

From Robin's travels

Like manoeuvring a jumbo jet at a city terminal, I watched as a couple used their two-way radio to back an eight-metre-long caravan in and out of their caravan park site. It seemed slightly ridiculous. On the other hand, a UHF radio is a useful piece of equipment. While travelling in outback country, north of Broken Hill, we stayed on a friend's station. Whenever we moved from our base camp, we were required to let the station know of our plans for the day, then inform them of our safe return. Our UHF radio proved perfect for this task.

Mail

Depending on the length of time you plan to be away, and family and neighbour relationships, the issue of mail can be tackled in various ways. By arrangement, the post office will hold your mail or forward it on for you to collect at a pre-arranged time, at a series of regional centres. Alternatively, you may have an arrangement with a family member or neighbour to collect your mail, and either sort it and forward important letters on to you, or simply hold it safely until your return. Uncollected mail – including junk mail – alerts thieves to your absence.

Mobile telephone service

The coverage that a mobile phone gives for country and outback use is the single most important feature to keep in mind when choosing a mobile for your travels. Telstra recommends a CDMA system for travellers. Telstra's mobile CDMA network is far superior to any GSM network. Its range Australia-wide is continually expanding. An external 'high-gain' CDMA antenna –

mounted on your vehicle's roof – greatly improves its ability to send and receive signals.

Satellite telephone service

You may choose to have a satellite-based tele-phone and facsimile service provided by Telstra, Optus or Vodaphone. This gives coverage across the whole of Australia, as well as some distance offshore.

Telephone answering service

A Telstra message bank or answering service is a simple and effective way to keep in touch. It enables you to recover messages from any tele-phone anywhere in Australia, and can be based on your home telephone or mobile phone num-ber, or by means of a number specially allocated for this purpose. You can choose to use your mobile phone for emergency purposes and for connecting you to an answering service – yet never have it disturb the serenity of your bush camp!

TV and radio

Travelling isn't about TV reception. It's about real life, in a real country, by a real you. On the other hand, some people find their travel experience goes more smoothly with a television. If commu-nication with the outside world is your only reason to include a TV set, consider a radio instead. Radio keeps travellers adequately informed and entertained, along with a supply of books, letters to write, and some games.

From Robin's travels

By listening to local radio, you'll find that the issues of the area become clear within a very short period of time. For instance, in the Mallee region of Victoria, extending to and including the Murray River, scarcity of water, salinity and a proposed toxic dump site were very much on the agenda when we travelled in the winter of 2004. Awareness of these issues helped us, as travellers, to connect with the locals in an environmentally sensitive way.

'Australia All Over' via ABC local radio

Every Sunday morning between 5.30 a.m. and 10 a.m. on ABC radio, the program 'Australia All Over' links together all Australians – but especially those on the road, or as Macca says, 'those on the wallaby'.

This program is a rich blend of talkback radio, letters and emails, country music, bush poetry and plenty of humour. Above all else, Macca's program hints at what being Australian is all about. People from all walks of life ring in and have their say: truckies, miners, grey nomads travelling in their caravans, people on cruising yachts, baby boomers on the 'wallaby', teenagers, children and senior citizens to name but a few. All relate firsthand experiences. Every now and then Macca goes on the road himself, and does a series of outside broadcasts and concerts. Some of these involve back-of-beyond celebrations and reunions. No matter where they are, these events are popular. 'Australia All Over' is a feel-good program, guaranteed to put a smile on your face.

Chapter 11
Bites, Stings and Nuisance Things

INSECT CONTROL

Wherever you travel, insects can make life difficult. By making use of a flyswat, flyscreens, sunshine, mosquito nets and a broom you will be practising green insect pest control.

The aim is to make insect pests unwelcome in your caravan, motorhome or campsite by cleaning away all food and water sources, and by disturbing breeding and resting places. Food scraps, dust, areas of moisture and dirty surfaces all encourage pests and must be eliminated.

Make sure that all food is stored in sealed containers, and that 'wet' areas are dry. Cockroaches and other pests need food and moisture. For detailed information about nuisance ants, bees, clothes moths, cockroaches, dust mites, silverfish and weevils see *Chemical-Free Pest Control* by Robin Stewart.

Repellents

An effective repellent will help when sandflies, march flies, bedbugs, leeches or ticks are a problem – applied to skin, clothing or both. When planning to swim, apply repellent both before and after being in the water. There are a large number of repellents on the market. Those containing the chemical DEET are recognised as the most effective. However, other commercial preparations such as natural pyrethrum-based repellents are available, as well as the more 'natural' repellents based on oils such as citronella and tea-tree. In general terms, the protection offered by 'natural' repellents is limited (30 to 60 minutes) and may give a false sense of security.

An effective repellent can be made up using an almond oil (or sorbolene) base with 10 to 15 percent of Dettol and a splash of citronella. Alternatively, use an olive oil (or grapeseed oil) base with 25 percent citronella, eucalyptus or tea-tree oil. A mixture of equal parts of methylated spirits, vinegar and eucalyptus oil can also be used as an effective repellent. Rub on your wrists, ankles, neck and forehead, but avoid contact with your eyes.

Treating bites

Depending on the type of bite and degree of irritation and pain, use any of the following to reduce the heat, itch and pain of the bite. You may need to experiment to find the one that works best for you.

• Place ice cubes in a tea towel or use a packet of frozen peas wrapped in a tea towel to take away some of the heat. Alternatively, plunge the bitten skin into a bucket or basin of cold water. Next, apply a thick paste of bicarb soda and cover with a cold wet cloth.

- Moisten bicarb soda with vinegar and apply.
- Smear with tea-tree, lavender, citronella or eucalyptus oil.
- Apply apple cider vinegar.
- Soothe with equal parts apple cider vinegar, eucalyptus oil and citronella oil.
- Smear with Vegemite.
- Soothe with equal parts cold tea, herbal vinegar and methylated spirits.
- If you're at the beach, try the juice from pigface to reduce the pain, itch and swelling. Simply collect some of the pigface creeper that grows wild in the sand dunes – the sort that has big purple flowers during the spring. Break open the fleshy leaves and squeeze the gel straight onto your insect bites. Cover the bites with a soothing layer of gel and then place cold wet towels on top. This is also a first aid treatment for sunburn.
- If you're in the bush, squeeze the juice from young bracken stems, or try the juice from a eucalypt (gum) leaf.

WILDLIFE

Many of the creatures in this section are native to Australia and are fully protected by law, such as snakes, sharks and crocodiles. There's no need to be frightened of these creatures. Nobody need be bitten or stung. In fact, there's every reason to admire and appreciate them as part of the web of life in and around Australia.

We need to remember that we're guests in the habitat of all native species; therefore, we must respect their territory and behave in a way that doesn't threaten them. Knowledge and understanding of the creature involved helps us do this in a sensitive manner, without injury or inconvenience.

Animal (dog, cat, horse, lizard, python) and human bites

Wash the bite under cold running water, then dry and apply tea-tree oil. Alternatively, sponge clean using a salt and water solution – half a teaspoon of salt dissolved in one cup of boiled water.

Due to the large population of harmful organisms that live in an animal's mouth (and beak and claws), any bite will be vulnerable to infection. It's wise to have a tetanus booster if bitten by an animal, and to keep a close eye on the wound. Pain, inflammation and oozing are suggestive of an infection, and should be checked by a doctor. Antibiotics will usually be required. A human bite also holds a high risk of infection, with the transfer of hepatitis B or C viruses a possible factor.

Ants

Bull ants (also called bulldog ants) and jumper ants (also known as 'jumping Jacks' or 'Jack jumpers')
This family of ants is aggressive and they cause considerable pain when they bite. Never annoy bull ants or jumper ants. Some people are susceptible to the poison injected by these ants, with possible fatal results. Following a sting, susceptible individuals should seek immediate medical attention. These people may need to carry appropriate medication at all times.

Unlike a bee, these ants are capable of stinging repeatedly. Unfortunately, too, bull ants in particular wander around at night as well as during the day; so when you're camping, barbecuing or walking in the bush, it pays to keep your eyes wide open and use a torch after dark.

From Robin's travels

On picking up some old bricks to construct a camp fireplace, I disturbed a nest of bull ants. It didn't take long before I realised I was in big trouble. If you've ever had the unpleasant experience of two or three bull ants climbing up the inside legs of your jeans, I'm sure you'll agree it can be an agonising experience. As well as causing much panic! Naturally, the removal of my jeans was top priority, regardless of amused spectators. Plunging my legs into the soothing coolness of a nearby creek became a matter of urgency as well.

Green tree ants (tropical)

These small ants are aggressive and have a very painful bite. Brushing against the leaves of tropical trees can cause you to accidentally collect these ants on your clothing – with painful results.

Bedbugs

Being bitten by the travel bug takes on new meaning when applied to the topic of bedbugs. Unfortunately there's been a resurgence worldwide of these unpleasant insects, even in the most civilised societies. The reason is probably the result of mass tourism into underdeveloped countries, with bedbugs hitching a ride back home in visitor's luggage, bedding, clothes and souvenirs. Garage and car boot sales may also have helped with their spread.

Bedbugs feed on human blood, are four to six millimetres long and have lentil-shaped bodies that are rusty brown in colour. They don't fly and have flattish bodies so they can hide in tiny cracks. Their eggs are laid in crevices and cracks in places such as timber bed frames, mattresses, suitcases, behind skirting boards and beneath wallpaper. The eggs hatch into tiny nymphs that undergo many moults before reaching adulthood.

These blood-sucking bugs are paper-thin when hungry, and bloated after a large meal of human blood. The peak of feeding occurs a few hours before dawn. After feeding, the bedbug retreats to its hiding place to digest its meal. They are able to survive for one year without food. The bite leaves an itchy lump, due largely to the bedbug's saliva. They are not known to transmit disease to humans, but they do cause sleepless nights! Normally nocturnal, they are attracted by the heat and smell of their host.

If you think there's a possibility that your luggage may be carrying bedbugs, wrap in black plastic and place in sunshine for three to four hours. Do the same with any suspect second-hand furniture, mattresses, clothing or bedding. Alternatively, pyrethrum spray or dusting powder may be used to kill bedbugs.

Hotels, motels and other establishments that offer sleeping accommodation can potentially harbour bedbugs. They thrive in centrally heated blocks of apartments and flats, especially when general and personal hygiene is low. So, what clues would make you suspect bedbugs?

- Assessing the general level of cleanliness as poor. Thorough vacuuming, washing with soap and hot water, and disinfection are essential means of control.

- Finding the discarded skins of developing nymphs on or between the sheets.
- Seeing tiny dots of excreta (bedbug droppings) on or between the sheets.
- Smelling a strong bedbug odour – sometimes described as musty or having a slight peppermint smell.

When choosing accommodation it's not as simple as avoiding the cheapest on offer. Budget accommodation may, in fact, be spotlessly clean, whereas more expensive accommodation may be poorly managed and contaminated with bedbugs! Some types of accommodation, for instance caravanpark cabins, require that guests bring their own bed linen. Alternatively, they hand out clean folded sheets.

Remember, as a paying guest you have the right to inspect your room prior to committing to stay and paying.

Black swans

The musical honking of swans at night is an eerie yet delightful sound, very much associated in my mind with water, a full moon and camping under the stars.

Black swans can, though, be aggressive when defending their nest – a large loosely woven structure made up of rushes and reeds, within a swamp or lake – or rearing their young. The male in particular will chase an intruder away from its cygnets, and is able to leap at and strike a person using its half-folded wings. The force of a strike is significant, enough to break a human arm or leg – even to knock an adult unconscious.

Rivers, billabongs, swamps and lakes are their favourite haunts, and late winter to spring their usual breeding time. Since most of us choose to camp beside water whenever possible, so the presence or otherwise of nesting swans needs to be considered. It's easy to avoid confrontation with these magnificent birds by:

- setting up camp beyond their breeding territory
- keeping your dog well away
- never feeding the swans
- avoiding standing between the parents and their young, backing them into a corner or invading their personal space if taking photographs of them.

Camels (One-humped)

Wild herds of camels (a legacy of the Afghan traders) can be a problem in some remote arid regions. Bull camels come into season (rut) from about April through to September. During the rutting season (the mating season), the bulls become extremely protective of their females. In fact, a feral bull camel can kill a person if that person happens to be in his way, or between him and one of his females. A bull camel, with his herd of females, can also block the track you're travelling along; a very intimidating situation. If faced with this scenario, close the windows of your vehicle sufficiently to prevent a camel's head invading your space, and remain in your vehicle. Eventually the camels will depart.

From Robin's travels

When travelling west of Alice Springs along a sandy track dotted with mulga scrub, our VW Kombi van was suddenly dwarfed by a bull camel uttering long, loud gurgling sounds. He stood over our vehicle, his long muscular neck extended, his teeth bared, with tongue lolling and frothy foam dribbling from his lips. But it was his spitting, aimed directly at my face, that made me feel violated.

A herd of about two-dozen females gathered behind him, grunting softly and uttering deep-throated moans. We sat quietly inside our vehicle until they departed, which took ten minutes or so. It was a long ten minutes. When we judged it safe, we stepped out onto the red sandy track, immediately aware of the musty smell of urine hanging heavy in the air. Dotted with camel footprints, the track snaked ahead and, as we watched, the mob wandered off into the shimmering distance.

Cane toads

The introduced cane toad is an example of biological control gone wrong. Large, unpleasant to look at, poisonous and expanding its range at an alarming rate, this creature is a serious pest. Cane toads have large swellings on each shoulder from which poison is squirted if the toad feels threatened or is handled roughly. If you've had contact with a cane toad, wash your skin very well using soap and warm water. If your dog has swallowed any of the toxin, rinse out its mouth using salty water, then force it to swallow salty water until vomiting occurs. Seek urgent veterinary help. Cane toad venom causes convulsions leading to paralysis and then death.

Cassowaries

The lush tropical rainforest of northern Queensland is home to the rare and endangered cassowary. This large, flightless bird is found around Mission Beach and the Daintree, and also on the Cape York Peninsula. These magnificent birds play a vital role in dispersing rainforest seeds.

With its uneven temper, helmet-like horny head, strong beak, long legs and powerful claws (three toes on each foot, with the nail of the inner toe elongated into a lethal knife) it's capable – if provoked – of charging, knocking over, pecking and kicking with fatal results. With a height of about two metres and about 60 kilograms of weight behind it, this rainforest warrior will terrorise anyone who dares take liberties with it.

If you catch sight of the bright blue neck and strange helmeted head of a cassowary, maintain eye contact, then back away very slowly, trying to keep a tree between you and the cassowary. Never approach, corner, chase or run away from a cassowary. Cassowaries can run much faster than humans. Cassowaries have a particular dislike of dogs, so keep your pet well away. Never attempt to feed a cassowary in order to take a photo.

Centipedes
Most centipedes are harmless. In tropical areas, though, the centipedes are larger and their bite more of a problem. A pair of fangs (containing venom glands) are located on the lower part of its head, and although the bite may hurt a lot, no deaths have been recorded. Disinfect the punctured area. Then, to help reduce the pain, soak the bite in very cold water, or apply an icepack.

Crocodiles (saltwater)
Saltwater crocodiles, sometimes called estuarine crocs or salties, live on the coastal fringe of tropical northern Australia – generally speaking, north

of Bundaberg in the east and Carnarvon in the west, meeting in the middle. They inhabit not only the coastline, but also many northern estuaries, freshwater coastal rivers, billabongs, small waterholes and creeks – and even bask on beaches. Salties can be found inland, lurking in freshwater billabongs and swamps, as far as 300 kilometres upstream. They've also been seen 100 kilometres out to sea.

Before you swim in any water in northern Australia, first check with the locals. Always obey crocodile warning signs as attacks from crocodiles in these areas are becoming more commonplace. Saltwater crocodiles may reach seven metres in length and weigh more than ten men. They are fearless, amazingly difficult to see, and can leap and then attack with extraordinary speed. As highly efficient hunters they are able to tackle humans, cattle, horses and goats. On the other hand, they are only dangerous to those who fail to respect their natural environment, disobey warning signs or behave in an irresponsible manner.

The world's largest predatory reptile prefers freshwater for many of its activities, including breeding. Its title of saltwater crocodile can be confusing. To prevent danger, while in crocodile territory:

- Obey warning signs.
- Never drink alcohol then swim in crocodile waters.
- Don't camp within 50 metres of water.
- Don't clean fish or leave food scraps near water.
- Don't wash your plates, cutlery and cooking utensils directly in waterways.

- Don't wade out into, or swim in waterholes or estuaries.
- Stand well back from water.
- Never lean out or trail your hand from a boat, especially in tidal estuaries.
- Remember that your dog will be a delicacy that a croc will find irresistible.
- Never place yourself between a crocodile and water.
- If chased by a croc, run in a zig-zag manner and fast. Then climb the nearest tree.

November is the start of the breeding season. Crocs are on the move, ahead of mating. Be particularly careful at this time of the year.

During the breeding season (November to March), the female saltwater crocodile selects a nesting site close to fresh water. Her nest-mound (designed to be above flood level) is made up of sticks and leaves, about two paces across and an arm-length deep. The mother crocodile guards her nest for the 90-day incubation period, behaving in a ferocious way if anyone approaches. Therefore:

- During the breeding season be aware of the breeding habits of this ancient creature and respect its need for solitude.
- If you accidentally stumble upon a nest-mound, move away fast, but cautiously.
- If you hear the squeaks and piping calls of hatchling crocodiles, stay well away!

Anyone mauled by a saltwater crocodile (or shark) will be severely traumatised and may be suffering appalling injuries. Immediate first aid is essential:

- Stop the bleeding.

- Clear airways, and resuscitate if breathing fails.
- Treat for shock.
- Seek urgent medical care as the risk of bacterial infection is high.

Crocodiles (freshwater)

The smaller freshwater crocodile is also fully protected by law, lives in fresh water and breeds at the end of the dry season, from October to November. They are commonly called freshies. They can become annoyed, but are usually not dangerous. Treat them with caution and do not disturb unnecessarily. On occasions, they have been known to attack, kill and eat dogs, and bite human legs. But left alone and respected as part of the ecology, they're not a problem – indeed, they're a privilege to see in their natural surroundings.

Dingoes and wild dogs

In many places throughout Australia dingoes have interbred with domestic hunting and cattle dogs. These feral crossbreeds (as well as pure dingoes) have become a problem in some popular tourist destinations such as Fraser Island and Uluru. Likewise, in the high country of New South Wales and Victoria, wild dogs have become a problem. In many of these areas the breeding of sheep and cattle has become impossible.

In alpine regions, bushwalkers and campers need to be aware that being confronted by wild dogs is becoming more frequent. Wild dogs have stalked and then bailed up hikers, snarling and barking in a threatening way. When wild dogs are in a pack and hungry, attacking adults and snatching young children is more likely, and has been recorded. For this reason it's unwise to camp or hike alone.

Normally wary of humans, dingoes won't usually attack unless cornered. They will, however, approach camps in the bush; partly to scavenge food (they are the ultimate opportunist feeder and predator), partly out of curiosity. Where the dingo has interbred with large aggressive hunting breeds, however, the likelihood of wild dogs attacking people has become more likely. For this reason:

- Never attempt to tame or feed dingoes or wild dogs, no matter how skinny they appear.
- Securely fasten the doorway of your tent, especially at night. Dingoes and wild dogs have been known to carry off food and clothing.
- Never follow a wild dog with a camera, thereby causing it to feel threatened, or back one into a corner or up against a barrier of some sort.
- Never let children or teenagers try to pat, tease or follow dingoes or wild dogs.
- Always leave your campsite free of all food scraps before retiring for the night.

If there are dingoes or wild dogs hanging around your campsite, don't allow children to play with loud noisy toys, argue, shout, hit, tease one another, make erratic movements and noises, or play-fight. These types of behaviour can precipitate an attack.

If a dingo or wild dog walks stiffly towards your camp, then freezes in its tracks, stares directly at you, growls, curls up its lips to expose its teeth, then raises the hair over its shoulders – realise you've got a big problem. Invasion of its territory, protection of a mate or litter of pups, or fear can provoke an attack. In this situation it's recommended that you:

- Remain calm.
- Stand tall and keep absolutely still, with your feet slightly apart and your back against something solid, if possible.
- Keep your eyes on the wild dog, but don't stare directly into its eyes. Staring can be threatening to any dog, wild or otherwise.
- If the wild dog rushes forward and butts its nose hard against your leg, keep standing completely still and don't react. Lack of reaction will confuse the dog and it will usually retreat.
- If knocked to the ground, pretend to be dead. Curl up into a tight ball, with your arms protecting your face. Keep very still. Remain like this until the wild dog departs, which it will eventually, through lack of reaction on your part. The more you struggle, hit, kick, push away, shout and yell, the more aggressive the dog will become.
- When the wild dog has lost interest, very slowly move towards your vehicle, but keep a careful watch on it without direct staring, and freeze if it becomes aggressive again.

Flies (bush flies, blowflies and march flies)

Flies prefer to live outside and can, therefore, be a problem when you're camping. They can make eating outside a real misery. One of the reasons we prefer not to share our food with them, or swallow them along with barbecued meat, is that bush flies are filthy in their habits, breeding in dead carcasses and faeces.

Bush flies find us attractive because we're a mobile source of food and moisture: sweat, tears, saliva and blood from wounds provide a bush fly with sustenance. They are very annoying through their sheer persistence and their habit of congregating around mouths, noses and eyes. They are often responsible for spreading diseases of the eye.

Blowflies are attracted by the smell of cooking meat, but also the odour of dead animal carcasses and every form of dung imaginable. They can be a big problem when cooking meat over an open fire or barbecue.

March flies are a problem from late summer through to autumn, depending on your area. When a march fly bites, it pierces your skin and sucks up the blood that flows. The bite hurts and can be itchy and sore for a number of days.

Before you reach for a spray can and spray yourself or fill the caravan with toxic vapours, remember that there's always a green solution. For example, if there's a fly in your caravan that's buzzing around and driving you crazy, don't reach for the spray and drown that single fly in pesticide. Simply turn off the light and open the door – with a light on outside – and the fly will depart. Alternatively, use a flyswat. Now you will be able to sleep peacefully, breathing air that is fresh, as well as completely free of flies.

Here are some simple green ways to reduce the impact of flies:

- Reduce their breeding grounds by burying all pet droppings and keeping rubbish well sealed within bins.
- Use flyscreens, flywire doors and fly nets (over food) to prevent flies being a nuisance.
- Use a flyswat to kill the occasional fly that beats the barriers.
- Do not leave uncovered food lying on benches or your sink, and clean up spilled food.

- When flies are particularly bad, use eucalyptus oil spray as a repellent on door and window frames. Vinegar wiped over surfaces also helps to repel flies.
- Place stems of rosemary on your barbecue. A rosemary plant in a pot is an excellent idea if you're a long-term traveller.
- Use a fly net over your hat, to keep annoying bush flies away from your face.
- Always brush your back before stepping indoors. In bush fly country, their habit of taking a free ride – in their hundreds – on people's backs can be a problem, especially if you want to go indoors.
- If you're eating outside, learn to eat with a fork or spoon only, so the other hand is free to shoo away bush flies.

Kangaroos

Normally, kangaroos are peace-loving, grazing animals that we are proud to display as our national symbol. There are occasions, however, where the male of the species becomes aggressive. These buck kangaroos are sometimes called 'rogue roos'.

Red and grey kangaroos, as well as wallaroos, can seriously injure a person. Rearing to a height of two metres and weighing about 80 kilograms, a red buck is able to grasp its victim with its forepaws, then rip with its powerful hind-leg toenails while balancing on its mighty tail. Scratches and bites to the victim's chest, head and neck can be life-threatening. A kangaroo has sharp incisor teeth that can bite through leather!

When camping in any park-like setting throughout Australia, it's a good idea to think of kangaroos and take sensible precautions. You can do this by following these ground rules:

- Take extra care when around kangaroos that are heat- or drought-stressed. In this situation, kangaroos are forced to graze close to country towns where grass is cut and watered. These roos will seem tamer than they actually are.
- Don't try to pat or feed these kangaroos, especially if there's a mature buck kangaroo nearby.
- If photographing kangaroos, avoid confronting a buck in particular with your camera. Kangaroos feel threatened if you come too close and point a camera at them.
- Never allow your dog to approach, bark at or chase kangaroos. Many a person has been injured through their dog being chased back to them. The next thing you know is that the kangaroo has turned its attention to you.
- Don't ever pat a hand-reared buck kangaroo that has been released into parkland or back into the wild. With no fear of humans, these former pets can be the most dangerous of all.

Any injury needs to be treated seriously, as the risk of bacterial infection is high.

From Robin's travels

For many years, rearing orphaned kangaroos was a passion of mine. Some joeys came to me as the result of road accidents – the mother dying, leaving behind a tiny joey in her pouch. Others were the result of shooting. I never could quite understand why somebody would shoot a female kangaroo, then drive all over the countryside trying to find someone to care for the joey.

Over the years I've cared for many other young animals, but kangaroo joeys have to head the list in terms of endearing characteristics. One grey buck joey was the most memorable: I called him Jarrah. Until he reached sexual maturity, Jarrah was the most loving and trustworthy creature imaginable. But all that changed within a very short period of time. Suddenly I was in danger of being ripped apart, especially when I had my monthly period. After much heartache, we relocated Jarrah to the Werribee Park Zoo, where he could live with others of his kind, in open parkland. Subsequently, one of the rangers at the Zoo told me that he'd got the fright of his life one day while constructing a fence, to discover Jarrah's enormous frame towering over him.

Leeches

Abundant in rainforests, leeches range from the tropics to the southernmost tip of Tasmania. Since bushwalkers tend to brush past ferns, long grass and rainforest foliage, they are most at risk of collecting leeches – especially around their ankles, necks and wrists.

If a leech is attached to your skin, sprinkle it with salt or vinegar. Alternatively, roll it off using your hand. Clean the wound with salty water and stop the bleeding using pressure. Relieve the itch with eucalyptus or tea-tree oil, very cold water, an icepack or bicarb soda moistened with vinegar.

Blood poisoning can develop as a direct result of a leech attaching itself to you. Disinfect the area thoroughly (using the above methods) and see a doctor if your lymph nodes swell or become painful.

Magpies

Magpies, kookaburras and masked lapwings (often called plovers) guard their nests and young by swooping anyone who strays into their territory. Defending eggs and young fledglings during springtime is instinctive behaviour that humans must learn to live with – and it does only last for about six weeks. After that time, these birds return to their happy selves, entertaining us with their melodious songs and amusing antics.

When selecting a campsite it's wise to avoid any swooping area, for your own and the birds' sake. It's unnerving to be the victim of swooping attacks from above and behind whenever stepping out of your caravan or tent.

If cycling is part of your travelling routine, always dismount and walk through the swoop zone. This will ensure that you don't have a nasty fall as a result of an unexpected dive-bombing bird. Wearing a helmet offers head protection from unexpected attacks.

Using their strong and very sharp bills, these birds can inflict painful wounds: a gash in the back of the neck, face, scalp or eye. If wounded, clean the injury and then see a doctor, as there's a high chance of infection.

Marine stingers
Marine Stinger Hotline: 08 8222 5116

When a person is stung by a jellyfish, a large number of tiny poisonous threads (invisible to the naked eye) are shot into the skin, where they break off while at the same time injecting poison. When in marine stinger territory – especially between the months of October and May – add a two-litre bottle of vinegar to your emergency kit.

Bluebottles (Portuguese man-of-war)
These common sea creatures, which are bright blue in colour, live all around Australia, sometimes

floating about in large shoals. More often, though, they're found in rock pools along the east coast. Although capable of inflicting very painful stings, their stings are usually non-fatal.

Wash off the tentacles using plenty of vinegar, then put an icepack over the sting to relieve pain. If you haven't got any vinegar, scrape the tentacles off the skin using a blunt piece of cardboard or a swimming flipper – but don't use your fingers. If numerous stings have been inflicted, nausea and difficulty in breathing may occur. If this happens, seek urgent medical attention.

Blue-ringed octopus
It's the electric-blue rings of this commonly found octopus that gives the animal its common name. These bright blue iridescent rings light up when the octopus feels threatened: a warning sign that the octopus is venomous and may inflict a deadly (but not necessarily painful) bite. They're most likely to be found living in rock pools. They occur all around Australia, with one species found only in northern Australia. About the size of a golf ball, they can sting through a wetsuit.

The first aid treatment is a pressure bandage and immobilisation – the same treatment as snakebite. Paralysis may occur, in which case artificial respiration will be required if breathing becomes difficult. Seek urgent medical attention, while at the same time reassuring the patient and keeping them warm and rested.

Box jellyfish (also known as sea wasps or marine stingers)
These small-bodied, pale blue or milky, semi-transparent jellyfish are found in tropical waters surrounding the northern coastline of Australia.

Able to deliver a fatal sting, these jellyfish are particularly active throughout the summer period, especially after local rain and when the sea is calm. It's considered too dangerous, therefore, to swim in northern waters between October and May. Always ask a local whether or not it's safe to swim in their particular locality.

The box jellyfish's 16 long, trailing tentacles (containing millions of stinging cells) wrap themselves around the victim. If the contact is extensive, the whip-like lesions on the skin will cause agonising pain, collapse and even death. Surfers may suffer minor stings from contact with tentacle fragments floating around in the water.

As a first aid measure, flood the area with vinegar for at least 30 seconds. Do not try to rub, wash, scrape or pick off the tentacles until unfired stings have been inactivated with vinegar. Ice or cold packs will help relieve the pain. Apply a pressure bandage over the sting and immobilise the limb. Keep the patient as still and calm as possible and seek urgent medical attention. Antivenene is available.

Conus (Cone) shells

Venomous conus shells are mainly found in tropical waters; however, there are southern species that are poisonous as well. When disturbed, a conus shell fires a venomous dart that inflicts a painful sting – sometimes fatal. Tropical conus shells are exceptionally beautiful and travellers enjoy finding them, especially when exploring coral reefs. Remember that it's illegal to remove live shellfish from most Australian waters.

Keep the patient as calm and still as possible, while at the same time applying a pressure bandage over the sting. Immobilise the injured limb, give artificial respiration if required and seek urgent medical attention.

Crown-of-thorns starfish
These starfish are considered a major threat to coral formations (especially on the Great Barrier Reef), because they feed on live coral. Every now and then their numbers build up to an alarming level. The crown-of-thorns starfish has many needle-sharp spines, each coated with an irritant mucous and venom cells. If a spine penetrates your skin, the pain can be intense. Strong footwear needs to be worn when exploring coral reefs, and protective clothing should be worn while swimming in places where crown-of-thorns starfish are common.

The first aid treatment is to carefully pull out any spines, bathe the wound in warm water and then apply an icepack to help relieve pain. Medical attention is recommended, though not urgently required.

Irukandji
These tiny, transparent jellyfish, armed with four stinging tentacles, are found in the warm waters surrounding the northern coastline of Australia. They commonly occur in swarms infesting northern beaches, usually for short periods of time. An unusual aspect of this jellyfish is that its sting doesn't hurt immediately. In fact, the venom takes about 30 minutes to have any effect. With time, though, excruciating and widespread pain is experienced, as well as many other distressing symptoms.

As a first aid measure, flood the area with vinegar. This will release the tentacles and inactivate any unfired stings. A pressure bandage needs to be wrapped firmly over the sting, then the limb immobilised. Keep the patient as still and calm as possible and seek urgent medical attention. To help relieve pain, apply an icepack or cold pack.

Stonefish

Stonefish are highly venomous – they have 26 poison glands – and can deliver a painful sting that may be fatal. The estuary stonefish lives on the muddy bottoms of bays and estuaries, ranging from northern New South Wales to Queensland, the Northern Territory and north-west Australia. The reef stonefish lives on reef flats and in lagoons around coral reefs. They are often found in very shallow water and are extremely hard to see, due to their excellent camouflage. They look exactly like an encrusted stone or a chunk of old coral. Their 13 venomous dorsal spines are each able to pierce the sole of runners! When walking on coral reefs, thick-soled runners should always be worn for protection against coral cuts, stonefish and cone shells. Turning over and picking up stones can be a hazardous occupation!

Emergency first aid involves bathing the sting with hot (though not scalding) water, to inactivate the venom. Then seek urgent medical attention. Antivenene is available.

Mosquitoes

Mosquitoes need to be treated seriously. They are the carriers of Ross River fever (occurring Australia-wide), Barmah Forest virus, Murray Valley encephalitis, yellow fever, dengue fever (occurring in Queensland and spread by daytime-feeding mosquitoes), malaria and many other parasites and viruses. The general rule is don't get bitten, and then you won't contract any of these debilitating mosquito-borne diseases.

In tropical and subtropical areas, mosquitoes are more of a problem between the end of the wet season and the beginning of the dry. In other areas mosquitoes multiply rapidly after heavy rain or floods, and favour brackish swamps, bill-abongs and anywhere else where water settles, such as areas where there is irrigation run-off. Mosquitoes can even breed in hoof prints.

Mosquito-proof screens offer protection during the hours of darkness. Cover up by wearing long-sleeved, long-legged, light-coloured clothing at dusk and at night, if outside. Be aware that mos-quitoes can bite through many light fabrics. Short sleeves, shorts and thongs leave you vulnerable to mosquito bites and many mosquito-borne viruses and parasites.

By burning mosquito coils (for instance, citronella coils), you will form a small area of protection in and around each coil. It's thought that highly scented perfumes, soaps and sprays make you more desirable to mosquitoes, so keep your use of these to a minimum. Other factors that affect your desirability or otherwise to mosquitoes include body odour and body heat.

Use a flyswat to kill the occasional mosquito. Use a mosquito net over your bed. If you sleep in a swag, a properly fitted mosquito net is essential.

Pigs (feral)
Feral pigs are a significant pest in Australia. They carry many diseases, such as brucellosis, Murray Valley encephalitis and leptospirosis. Feral pigs also cause damage to cereal crops and pastures. Dust and mud wallows, and up-rooted vegetation,

suggest strongly that feral pigs are in the locality. Feral pigs are likely to charge if cornered, so care needs to be taken to avoid a confrontation with a boar (armed with dangerous tusks), or a sow with piglets. Feral pigs are most active from late afternoon until a few hours after daybreak. They spend the warmer parts of the day resting in dust wallows, usually in groups.

From Robin's travels

While camping in central Queensland I had an encounter with a wild boar I'll never forget. Walking through some mulga scrub in the late afternoon with my labrador dog running ahead, I heard an aggressive snorting sound and looked up to see my dog running for his life towards me, with a black boar chasing him and rapidly gaining ground. Frantically I looked around for a tree to climb, but there was nothing – only mulga scrub too low and too flimsy to offer any real protection. I could see the tusks and yellow teeth by now, so I crouched behind some scrub, sweating profusely. My dog continued his gallop towards me, tongue lolling, and finally flung himself into my arms. I stood to my full height, while at the same time ordering my dog to 'sit and stay'. The boar stopped in his tracks. Remaining quiet and very still, with heart thudding, I watched as the boar decided retreat was, perhaps, the best option.

Pythons

Pythons are not venomous, but they do bite. If large enough, a python can squeeze the breath out of a person to the extent of causing death, but this is rare. An amethystine (scrub) python grows to about 3.5 metres and weighs approximately

20 kilograms. Its constricting coils can tackle anything from a wallaby to a calf to a young child, literally crushing their prey to death. Pythons frequently bask on tree branches and are very well camouflaged. Bushwalkers are advised to check tree branches before grabbing them for balance or support. Diamond and carpet pythons also mature into large powerful snakes.

A python's bite is likely to be contaminated with bacteria, so disinfect the wound thoroughly. It's wise to visit a doctor, as you may need antibiotics and a tetanus booster as well.

From Robin's travels

While living on King Island, we agreed to meet my brother (from Cooktown) at a caravan park in Sydney, where we'd booked an on-site van. What I hadn't counted on, however, was my brother bringing along a diamond python he'd rescued along the way.

So we were introduced to Tiffany, the diamond python, and I had to admit she was a stunning creature – from a distance. With the late afternoon sunshine highlighting her diamond-patterned skin etched in black and gold, she rested in a contented coil on the top bunk of the caravan. Unblinkingly, her beady eyes soon out-stared me, and I was forced to look away and ask my long-haired, bearded brother, 'Where will she sleep?'

After some negotiation, a compromise was reached whereby Tiffany would roam free in the caravan except when we were in bed. Then she would be confined to her pillowcase on the top bunk, with the end securely tied. The top bunk was as far away from the double bed as it was possible to get.

My brother of the northern rainforests handled this python as I would an orphaned kangaroo joey. Lifting her gently and speaking in a crooning voice, he allowed the two-metre snake to wrap herself loosely around his arm and shoulder, tasting his breath and skin with her long, forked tongue. As we unpacked, Tiffany flowed silently over the two top bunks, raising and lowering her head to check her surroundings. Finally she settled in a far corner, her head resting on a coil as she surveyed our preparations for the evening meal.

Several days later – after one night-time escape, when Tiffany found her way into Doug's backpack, but thankfully not our bed – we released her in a location where we felt sure of her safety and wellbeing. My brother was certain I'd benefited from the experience, and in future would feel more relaxed around pythons. I hadn't and didn't!

Sandflies (biting midges)

Avoidance is the best defence against sandflies, with estuaries, sandy river banks, islands, coastal scrub, lakes, swampy areas, brackish water, dense tea-tree and mangroves frequently the haunt of huge swarms of sandflies. Generally speaking, these are not good places to camp. If you find yourself in a sandfly area, protective clothing is a must, including on your head. Bites on the scalp can cause severe irritation, and may develop into oozing sores if scratched. Depending on your sensitivity, sandfly bites can either drive you crazy (to the point of delirium) with their itch or be of little consequence. If badly affected, it's sensible to use antihistamine tablets. To reduce the terrible itch, have a cold shower, then apply an icepack. See pages 180–1 for more hints.

Scorpions

Scorpions are found all over Australia, with the largest species up to 12 centimetres in length. Australian scorpions are not, however, as much of a problem as those found overseas. On the other hand, the further you travel north, the bigger and more poisonous the scorpions become. They are commonly found resting beneath small rocks in the bush; therefore, when you're fossicking around collecting stones with which to construct a fireplace, keep your eyes open and perhaps wear protective gloves. A scorpion's sting is located on the tip of its tail.

In humans, the poison can produce a burning or throbbing pain that may last for many hours. Very cold water or an icepack give relief from a painful sting. If allergic-type symptoms develop, see a doctor.

Sharks

Along the New South Wales and Queensland coastline, canal-front real estate is proving popular with both tourists and homeowners. Unfortunately, a problem has developed with bull whaler sharks. These dangerous and very aggressive sharks – nicknamed the pit bulls of the shark world – have attacked and killed people, as well as dogs. For this reason, it's considered unsafe to swim in any inland waterway connected to the sea. This includes a vast network of interconnected, man-made canals and lakes. Bull whaler sharks (about 3.4 metres long) are now known to penetrate deep into freshwater river systems and seem to seek out shallow inshore habitats.

On ocean beaches, a number of strategies are in place to protect people from shark attack. These include: shark spotter planes that fly over popular swimming and surfing beaches; lifeguards with lookout towers, who watch for sharks at popular swimming beaches, and mark out safe places to swim between flags; and beach-meshing of popular swimming and surfing beaches. Unfortunately, this protective netting traps other fish and marine mammals such as dolphins.

The great white shark (white pointer) is the largest and most dangerous shark. It can grow up to six metres long. Nine shark species, including the great white shark, are fully protected in Australian waters.

To help prevent an attack, follow these easy guidelines:

- Never swim near or take a small boat close to a seal colony. White pointers are common visitors.
- Avoid wearing a black wetsuit – a shark may mistake you for a seal, one of their favourite meals.
- Never swim in water that contains blood; for example, a place where fish have recently been cleaned. A shark can detect one part blood in 10 billion parts of water.
- Even knee-deep water can be unsafe.
- Never swim alone, especially in surf.
- Always obey signs that warn swimmers of the danger they may face from sharks.
- Avoid swimming at night, in the late afternoon or on an overcast day.
- Never swim in a river estuary at dusk.
- Avoid swimming in discoloured water.
- Don't swim where others are fishing.
- Avoid swimming with your dog – the dog's odour is very attractive to sharks.

- Do not wear shiny jewellery, as this may attract a shark's attention.
- Surfboards and small boats are sometimes attacked by a shark. It's likely that the shark has mistaken the craft for a seal or a large fish.
- If you're stranded in the ocean waiting to be rescued, hunch yourself up so that your arms and legs are close to your body. If in a group, huddle together with as few arms and legs extended as possible. Avoid splashing about like an injured fish.
- If you're a diver with a passion for underwater photography, take care not to corner or invade a shark's territory, or threaten a shark by pointing your camera at it. Defending territory can be the trigger for a shark attack.

From Robin's travels

I'd always been taught that swimming in a river estuary at dusk was risky, in terms of shark attack. Yet, when planning to set a net across the neck of a small bay one still balmy evening, sharks were not foremost in my mind. Each of us had a task: the women unrolled the net from its bag, waded out into shallow water then 'fed' the net to the men who waded out into deeper water – placing the net in an arc across the mouth of the estuary. The aim was to trap 20 or so salmon.

All of a sudden, when I was up to my waist in water and about 70 metres from shore, I heard one of the men yell, 'Shark!' Slicing through the bay moved a large dark grey fin, advancing stealthily in my direction. Then it disappeared. Painfully aware of my bare legs and the fading light I scanned the water for a fin. Where was it now? It surfaced nearby, its three-metre-long body thrashing about as it fed in a frenzied manner on the school of salmon. Then it dived again. In what direction? I wondered, my heart hammering against my chest, my mouth dry.

All four of us stood perfectly still, for we knew that sharks were attracted to noise, splashing and panicky movements. Gruesome images flashed through my mind, as seconds hovered like hours – of razor-sharp teeth ripping the flesh from my legs, of snapping jaws, sinister grey fins and blood-stained water. Yet, within minutes we had the bay to ourselves again, and could pull the net, which of course contained absolutely nothing except bull kelp and stones. We were foolish as well as very lucky.

Silver gulls (commonly called seagulls)
In popular areas, be aware that seagulls commonly snatch food from fingers – and even swoop to peck chips from the mouths of children.

Snakes (land)
All Australian snakes are fully protected by law. Most are shy and will slither away when they feel the vibration of your footsteps – especially if you deliberately walk heavily. Snakes tend to be more aggressive during the breeding season, which usually occurs from late October through to December. When picking up firewood in particular, care needs to be taken, as most Australian snakes are highly venomous. Therefore, it's a good idea to wear protective gloves when collecting firewood, and to be particularly cautious about handling hollow logs.

Wearing long trousers, socks and strong shoes is a good idea when walking in the bush – ideally gaiters in high-risk areas, and never thongs, bare feet or sandals. When dressed this way, you can spend time in snake country without fear of snakebite.

Dr David Fleay, one of Australia's leading naturalists and the first person to milk a taipan, always wore two pairs of thick socks. No snake could bite through this barrier, he said.

If you're sleeping in a tent, always check your sleeping bag, towel and clothes before using them, in case of unexpected visitors. Never prod at a snake with a stick, and never step over a snake. Always look carefully before stepping over logs – there may be a snake sleeping in the sun on the other side. Most snakes are active on warm summer nights; therefore always use a torch when walking around your campsite.

To treat snakebite, lie the person down and, keeping the patient as still and calm as possible, bandage firmly over the bite using an elastic or crepe bandage. If you haven't got a bandage, any flexible material will do; for instance, you can tear up a T-shirt or sheet into strips. A second bandage should then be used to apply a firm, even pressure over the entire limb. Bandage as firmly as you would bandage a sprained ankle. Now immobilise the limb by splinting as for a fracture. Note the time and try to get a description of the snake. Dial 000 for an ambulance. Keep the patient as calm and rested as possible.

From Robin's travels

With many scary snake stories stamped in my memory, I'm very aware of snakes and careful in my choice of clothing and behaviour when in their habitat. I admire the beauty of their skin, their graceful way of moving and place within the Australian environment; however, I'm not entirely relaxed around them.

One memorable occasion occurred on King Island in Bass Strait, while camped at a cliff-top muttonbird rookery. Concentrating more on conversation than where I placed my feet, I trod on a black tiger snake which was sunning itself in the protection of a tussock. In an attempt to free itself from my foot, it struggled, flattened its head and neck, hissed aggressively and struck at the back of my jeans. As it raised its body to strike a second time, I bolted towards the cliff-face. We were two hours away from a hospital, and tiger snakes are known to possess large fangs capable of injecting liberal quantities of venom. As they possess one of the most potent poisons in the world, I was not keen on being bitten, as even a small amount would cause death unless antivenene was given promptly. With its tiny eyes flashing, it gave a hacking cough and slithered after me at great speed. Well over 1.5 metres in length, it caught up with me and struck at my jeans with the force of a whip. Stumbling over stooped herbage and rushing through a clump of sword grass, I felt it strike again and again, its glossy black skin propelled like lightning through the salt- and wind-stunted vegetation. It reared and struck one more time, then slithered beneath a salt-bush. Jeans and sturdy elastic-sided boots had saved the day.

Snakes (sea)

Many sea snake species live in Australia's tropical waters, and some have powerful venom. Most are readily recognised by their paddle-like tail. Antivenene is available. The pressure-immobilisation method of treatment is the same as for land snakes. Urgent medical care needs to be sought.

Spiders (Sydney funnel-web spiders, mouse spiders, red-back spiders and white-tailed spiders)

By following these simple hints, you'll greatly reduce the chance of being bitten by a spider:

- Avoid walking outside with bare feet.
- If you need to get up during the night, wear slippers or shoes, and either turn on a light or use a torch.
- Always wear protective clothing and gloves when gathering wood for your campfire and when picking up rubbish. A spider usually releases its venom before it strikes, which means that when it strikes through clothing, the venom is wiped off the fang before the fang strikes flesh. Although it will often bite repeatedly, only a single dose of venom is released.
- When using an outside camp toilet at night, take a torch and inspect the seat before you sit down. Jokes about being bitten on the bottom while sitting on a toilet seat are usually not amusing to the victim!
- Don't leave bedding, towels or clothes lying on the floor, especially overnight. If you do, shake them very well before using them again. Check your bed before retiring.

Sydney funnel-web spiders

This spider is usually found within a radius of about 160 kilometres from the centre of Sydney. Some authorities believe its range extends further up the coast and also into central New South Wales, as well as into south-east Victoria.

Funnel-webs are large, glossy, blue-black spiders with heavy bodies covered in fine velvety hair and relatively short legs. The female is about 35 millimetres long, and the male 25 millimetres long. Sydney funnel-webs are nocturnal and don't like being in sunlight, heat or dryness, preferring to shelter beneath fallen logs and rocks in lush valleys, in moist soil beneath buildings, in crevices in rockeries or in burrows in compost.

When this spider bites, it rears up with its fangs raised while it braces itself against the ground. The large fangs strike downwards like those of a snake, with the venom designed to immobilise prey. Two clear puncture marks will be visible.

This intensely painful bite leaves its victim feeling extremely distressed and frightened. What should you do in the event of a bite?

- Pressure-immobilisation first aid treatment is the same as for snakebite. Firmly bandage (as you would a sprain) the length of the bitten limb, then splint, keeping the patient very still. This will stop the flow of lymph, but not stop the arterial pulse.
- Keep the patient warm and calm.
- Seek urgent medical attention, as antivenene treatment usually results in a full recovery. If left untreated, a Sydney funnel-web's bite usually ends in respiratory or heart failure.

Australia has 35 funnel-web species. All funnel-webs are potentially poisonous to humans and should be treated with great caution.

Mouse spiders (*Missulena*) are also highly venomous, so their bites need to be treated in the same way as funnel-web bites. These squat, heavily built spiders wander around during the day and are found all over Australia except Tasmania. Sydney funnel-web antivenene is sometimes used to treat a severe mouse spider bite.

Red-back spiders

Red-backs live all over Australia, including Tasmania and the Simpson Desert.

Only the female red-back's bite is potentially fatal. The bite may be intensely painful at first, yet this is not necessarily so. The venom is very slow-acting. Sometimes it's only painful hours later and frequently affects other parts of the body, the nervous system in particular. Follow these steps to treat a bite:

- Wash the wound with soap and hot water, or salt and water.
- Apply an ice pack (ice or a packet of frozen peas wrapped in a tea towel, for example) to relieve the pain.
- Do not apply a pressure-immobilisation bandage and splint. Only a tiny amount of venom is injected and it moves very slowly throughout the body.
- Go to your nearest hospital or doctor; however, do not panic, remembering that the venom moves very slowly. Antivenene became available in 1956 and is highly effective.
- Keep the patient still, comfortable, calm and warm.

Never place ice directly on the skin. You can damage the tissue in the same way as frostbite. Simply wrap the ice in a tea towel or equivalent.

White-tailed spiders

These spiders are found throughout Australia, including Tasmania, and also New Zealand. It's quite common to find them hidden in bedding; in clothes left lying on the floor; in wardrobes; in curtain creases; or hidden in the folds of towels. If you get up in the middle of the night, you may see one running across the floor or across a wall.

White-tailed spiders are not aggressive and only bite if provoked, frightened or threatened. They prefer to retreat. As soon as you know you've been bitten, follow these hints to help minimise the risk of serious skin damage:

• Scrub the area with soap and hot water using a very stiff brush. The aim is to get to rid of all the surface venom and bacteria.
• Use an ice pack (or a packet of frozen peas wrapped in a towel) to reduce pain.
• Use a twice-daily application of manuka honey to kill any bacteria left on the skin and promote healing.

Other spider bites

If bitten by a black house spider, green jumping spider, wolf spider or an introduced fiddle-back spider, treat the bites of these species in the same way as a white-tailed spider bite.

Stinging trees and shrubs

There are four native species of the nettle family. Found in the rainforests of northern Australia, and usually growing on disturbed ground, these plants often have heart-shaped leaves. Their leaves and twigs are armed with fine hollow hairs, with a reservoir of poison at their base. When accidentally brushed against, the tips of the hairs break off as they penetrate the skin, injecting poisonous sap. This poison causes itching, pain, a rash, swelling and red spots to develop – even making victims delirious. Severe, long-lasting pain may result.

A remedy that locals use is depilatory wax. Melt the wax, allow it to cool to skin temperature, then smooth over the stung area. When the wax is cold and dry, peel off. Most of the stinging hairs will come away with the wax. Repeat if necessary. This will save many months of pain and irritation. For quick relief, use the wax as soon as possible.

Alternatively, cunjevoi lily, which often grows alongside the stinging tree, contains a juice that will help relieve the pain. So too will ordinary household vinegar. Cold water or an icepack help relieve discomfort.

Ticks
Ticks are parasitic creatures – in the same family as spiders, scorpions and mites – that pierce their host's skin and then suck up blood. Scrub ticks (commonly called paralysis ticks), which occur along the eastern coast of Australia as well as the north, can be lethal to young children and family pets. Bandicoots are the natural host of the scrub tick. It's wise to avoid walking through long grass, rainforest or bushy vegetation, especially in bandicoot country. Likewise, avoid sitting on the ground.

The first sign that a tick has attached itself is skin irritation that develops into a raised lump. On examination, this lump is seen to contain a tick. It's important to inspect young children, dogs and cats daily, especially in skin crevices, behind ears, in the hair and groin area, in neck folds, and under arms and around waistlines.

When people are outside in tick country, wearing light-coloured clothing with long-sleeves and pants tucked into socks reduces the likelihood of ticks attaching themselves. A light spray of eucalyptus oil will help repel ticks.

To remove a tick, follow these hints:

- To reduce the chance of infection, wash your hands and the area around the tick with an antiseptic.
- Using clean fingernails, curved scissors or tweezers, grasp the tick as close to the skin as possible.
- Gently lever the tick out, using a slight to-and-fro action. Do not twist or jerk the tick, as this will cause more of the venom to be injected.
- If it breaks off, leaving behind some tiny black pieces, don't worry. Instead, bathe with antiseptic and apply a poultice. For more information on poultices, see pages 147–8.

Ticks carry a range of bacteria and viruses, including Lyme disease, a debilitating bacterial blood infection that requires medical attention. Therefore, keep the tick taped to a card using clear plastic tape. Record the date and the part of your body bitten. If a rash or ill health develops – even up to one year later – the tick can then be properly identified.

From Robin's travels

Travelling with a labrador dog was perfect until we reached the coast of far north Queensland, where a scrub tick attached itself to our dog. Recognising the typical hind-leg paralysis, we knew exactly what was wrong but couldn't find the tick. By this stage, though, we were inland, and far removed from a vet. A vet did suggest, by phone, that we search in our dog's ears and mouth – and that's where we found the tick, embedded in our dog's lip folds.

By this stage, the paralysis was affecting his front legs and also vital organs such as his lungs. We nearly lost our much-loved dog, but several weeks of tender loving care restored his health and vitality. While he convalesced, we camped by a river near Normanton; a river well stocked with freshwater crocodiles, as well as swarms of mosquitoes and sandflies.

Wasps (European)

The European wasp (introduced into Australia in the late 1980s) is a dangerous pest with a savage sting. A single wasp can attack and sting repeatedly, and it encourages its friends to attack as well. European wasps do not like being disturbed and can sustain an attack for five minutes. Most attacks occur within seven metres of their nest.

European wasps are especially attracted to sweet drinks. Drinking directly from an open drink can or bottle is potentially very dangerous. A wasp may climb into the can and sting your mouth or throat when swallowed, causing you to choke and suffer severe difficulty in breathing. Using a straw or glass when drinking outdoors is a good idea. A glass will let you see what you're drinking; a straw will be too small to allow the wasp to pass through.

Outside barbecues and picnics attract European wasps, and their favourite foods include sandwiches, barbecued meat, strawberries, salads, soft drinks and beer. If a European wasp approaches your picnic, it's not a good idea to swat it, as a moving target will attract and provoke it further. The best idea is to stay still. If a European wasp has entered your caravan, kill it with a flyswat or broom. For more than one wasp, open an exit door, then sit still and keep a cautious eye on them. In time, they'll be attracted outside again.

Some people are hypersensitive to the toxins the wasp injects. These people require pressure-immobilisation treatment and urgent medical care. A sting on the soft tissue of the mouth or throat is potentially life threatening.

If you have an ordinary reaction (a hot red swollen mark, several centimetres across, with fiery pain), it's sufficient to dab the bite with vinegar or lemon juice – after you've moved indoors, of course. The sting, being alkaline, is neutralised with an acid. For very painful bites, use an icepack or soak the affected limb in a container of iced water. A cooler or icebox is usually large enough and will keep the water cold.

Further Reading

Julie and Enzo Coco, *Dogs on Holiday*, Go North, Queensland, 1994.

Joanne Colak, *Pets Welcome Accommodation Guide*, Gary Allen Pty Ltd, Smithfield, 2002.

Peter and Sandy Dennis, *Camping and Holidaying with Dogs*, Life. Be in it, Melbourne, 1998.

Flat Earth Mapping Pty Ltd, *Bush Camping with Dogs*, South Australia, 2003.

Kate Harte, *Holidaying with Cats*, Life. Be in it, Melbourne, 2000.

Lloyd Junor, *Caravanning Outback Australia*, Aussie Outback Publishing, 2000.

Craig Lewis and Cathy Savage, *Camping in Victoria*, Boiling Billy Publications, New South Wales, 1998.

Craig Lewis and Cathy Savage, *Camping Guide to New South Wales*, Boiling Billy Publications, New South Wales, 1999.

Chris and Yvonne McMaughlin, *The Rivers and Lakes of New South Wales: Canoeing and Camping Guide*, MacStyle Media, Victoria, 1998.

Chris and Yvonne McMaughlin, *The Rivers and Lakes of Victoria: Canoeing and Camping Guide*, Riverside Publications, Victoria, 1999.

Philip Procter, *Camps Australia Wide*, Pre-Press, Queensland, 2003.

Collyn Rivers, *The Campervan and Motorhome Book*, Collyn Rivers, Broome, Western Australia, 2003.

Collyn Rivers, *Motorhome Electrics and Caravans Too*, Collyn Rivers, Broome, Western Australia, 2003.

Collyn Rivers, *Solar That Really Works*, caravan and motorhome editions, Collyn Rivers, Broome, Western Australia, 2004.

Paul Smedley, *Bush Camps and Rest Areas across Australia: Australia's Major Inland Highways*, Highwayman Publications, South Australia, 1999.

Paul Smedley, *Bush Camps and Rest Areas around Australia: Highway One and the Stuart Highway*, Highwayman Publications, South Australia, 1999.

Paul Smedley, *Bush Camps and Rest Areas across Australia: Minor Highways, Byways and Tracks Throughout Australia*, Highwayman Publications, South Australia, 1999.

The following internet sites provide travellers with valuable up-to-date information relating to sustainable travel:

- Aussie Outback Tours <www.aussieoutback.com.au>
- Caravan and Motorhome Books <www.caravanandmotorhomebooks.com>
- Australian Government Department of Transport and Regional Services, Australian Greenhouse Office <www.greenvehicleguide.gov.au>
- Ecotourism Australia <www.ecotourism.org.au>
- Victorian Tourism Operators Association, Nature and Ecotourism Accreditation Program <www.vtoa.asn.au>
- AAA Tourism <www.aaatourism.com.au>
- YHA Australia <www.yha.com.au/australia/environment.cfm>
- Forestry Tasmania <www.tasforestrytourism.com.au>
- Discover Tasmania <www.discovertasmania.com>
- Earthwatch Institute <www.earthwatch.org>
- Kangaroo Island Secrets Visitor Guide <www.tourkangarooisland.com.au>
- Bogong Horseback Adventures <www.bogong-horse.com.au>
- Healing Dreams Retreat <www.healingdreams.com.au>

Index